CATS

Nature and Nurture

CATS
Nature and Nurture

ANDY HIRSCH

:01
First Second
New York

Drawn in Manga Studio EX 5. Colored in Adobe Photoshop CS5. Lettered with Comicrazy font from Comicraft.

Published by First Second
First Second is an imprint of Roaring Brook Press,
a division of Holtzbrinck Publishing Holdings Limited Partnership
120 Broadway, New York, NY 10271

Library of Congress Control Number: 2018944913

Paperback ISBN: 978-1-250-14312-9
Hardcover ISBN: 978-1-250-14313-6

First edition, 2019
Edited by Dave Roman
Cat Consultant: Wai-Ming Wong, PhD
Book design by Chris Dickey

Printed in China by Toppan Leefung Printing Ltd., Dongguan City, Guangdong Province
Paperback: 10 9 8 7 6 5 4 3 2 1
Hardcover: 10 9 8 7 6 5 4 3 2 1

Nowadays, it's all about the cat. There are currently more cats living in households than any other pet (except for fish). But despite cats' popularity, society seems to have a love-hate relationship with them, and it's been that way ever since humans and cats first met thousands of years ago! From being both worshipped *and* sacrificed by ancient Egyptians, to their association with witchcraft in the Middle Ages, to being considered either good *or* bad luck (depending on where you live), cats have experienced plenty of ups and downs during their relationship with us humans. Even now in the heyday of Grumpy Cat, Lil Bub, cat cafés, and over 26 billion views of cat videos on YouTube, not everyone loves cats—cats are accused of being mass killers of songbirds and vectors for disease, and a fair amount of people are just plain afraid of cats (a condition called *ailurophobia*).

For those of us who do love cats, there's nothing as soothing as the sound of a cat purring, especially if they're snuggled on your lap. But despite this intimate relationship, there's also nothing as *confusing* to many humans as their beloved cat. What is she thinking? Does he *really* love me? Why does he pounce on my head at 5 A.M.? Why does she seem to like petting and then suddenly bite me? And how do cats just seem to know where we want them to go to the bathroom?!

I had plenty of questions about the nature of cats, but it wasn't until one of my own passed away that my life changed forever: I decided I needed more cats in my life and walked into the local animal shelter to volunteer. Because I'd lived with cats for most of my life, I thought I knew all about them. But after my first few weeks in the shelter, I realized I still had a lot to learn. As territorial animals, cats form an attachment to where they live before they can form attachments to people. What was fascinating to me was how individual each cat's response to being in this new environment was—some cats were terrified, while others adjusted quickly. Were these personality differences caused by their genes? By how they were raised? And did they only come to light when cats were in a stressful environment like an animal shelter? These questions made me realize that I wanted to study cats and know everything there is to know about them!

Fast-forward to many years later, where as a scientist and a cat behavior consultant, my job is to help us *all* better understand cats. Cat owners come to me with all kinds of questions about why their cats don't get along, or why their cat won't use the litter box, or why their cat is keeping them awake all night. And my current research is trying to help us better understand the social lives and development of cats.

But if you want to understand cats, *really* understand them, there's just one thing you need to know. How they came to live with humans, what their

bodies are designed to do, and what motivates so many of their behaviors like playing, eating, and sleeping are all shaped by one thing: HUNTING.

Once you know how important being a predator is for cats, it opens up a whole new world. You won't see playing with your cat in the same way ever again! As you'll learn when you read about Bean, an itty-bitty kitty who starts this tale with a very empty tummy—there's only one way to satisfy that empty tummy—by hunting (or by having a human provide you with a bowl of food). Cats basically have two choices these days: be a killer or be a companion (the occasional kitty gets to do both). But the fuzzy companions who live inside still have killer instincts.

Unlike dogs, who leapt into domestication with a boundless enthusiasm, cats have tiptoed their way into our lives. The really cool thing about cats is that during the process of domestication, we didn't ask them to change much. We liked cats because they were warm and cuddly and caught mice, and they liked us because...well, probably because we were warm and cuddly and gave them food. Kind of the same reasons they like us now!

We didn't ask cats to change, but we changed their environment almost overnight! They went from living free in the streets to spending most of their time in an indoor environment—an environment that is no doubt safer for cats (and birds!)—but without some effort, might be a bit boring. After all, there's

not much to hunt inside, except maybe the occasional bug. As you'll see as you read ahead, Bean's human knows just what to do to turn the indoors into a kitty paradise—and you can do the same with your own cat!

Dogs wear their hearts on their sleeves; in comparison, cats may seem like a bit of a mystery. But I predict that is all about to change. In recent years, the level of scientific interest in cats has mirrored cultural interest in cats: in other words, it's been *off the charts*. Okay, maybe not like 26-billion-views-of-cat-videos-on-YouTube off the charts, but I'd say we are entering a Renaissance period for feline science. We are also recognizing that to properly study cats, we must be a little creative, like using technology (such as cameras and accelerometers) and observing cats in familiar environments instead of a laboratory (since they are so territorial) to get a better sense of what they really do.

So the book that is in your hands is a great place to start if you are interested in science and especially if you are interested in cats. You're going to leave with an understanding of how cats came to live with and love humans, and why we love them so. If you have a cat in your home, this book will help you be a better friend to them—and who knows, perhaps some of you will go on to study cats yourselves in the future. I hope you are inspired by this book and enjoy reading it as much as I did!

Dr. Mikel Maria Delgado, PhD
Postdoctoral researcher at the University of California, Davis, School of Veterinary Medicine, and cat behavior consultant at Feline Minds

WOOOOSH

3
2
...
CLAP
STAR TALES
EP. 901
PROD. CO. SHOUT

MRRRMPH
BEAN: 1 TRILLION VIEWS AND COUNTING

mrmph

Oh, did I doze off? Where's my sunbeam?

There!

Sorry, this lifestyle really takes it out of you.
YAWN!

Right, what was your question again?
"Can you describe the moment you became a star?"

Hmm, how long have you got?
You'll need some background to really appreciate this.

RAGIC
ORI G 1N
$3 99
Let me start at the beginning...

The first thing I remember is being *hungry*.
chirp chirp
The second thing I remember is wanting to *fix* that.
I didn't know where I was or how I got there. I was just a lone li'l kitty with an empty li'l tummy.
Keep still, you!
FREE KITTEN
What a world.

Think you're so *smart*...
chirp
...we'll *see* how smart you are...
chirp
...when you're chirping in my *belly*...
chirp
POUNCE
Hold it!
This right here? This tells you more about me than you might guess.

Oh, yes, look past my adorable face. This sweet smile that's charmed millions?

It strikes *fear* into prey around the world!

It's the smile of a *carnivore,* a meat-eater! Not just *any* carnivore, though—
Your meat, miss.
BEAN'S FAVORITE MEALS

a *hyper-carnivore!*
And your sides.
BEAN'S FAVORITE MEATS

'at 'eef're~
GULP
Ahem. Cat teeth are meant for one thing: eating meat. If you compare our teeth to other members of the carnivore club, you can really see our priorities.
Those big, sharp teeth toward the back of dogs' mouths are the *carnassials*. They work like scissors to slice soft flesh. You know, meat.
Behind them are *molars* for chewing. They've got flatter tops that are great for grinding up plants! Dogs'll eat anything, and their teeth show it.
gnaw gnaw SLOBBER
CRUNCHA MUNCHA
Bears' teeth are much better suited for grinding than slicing. They may be part of the carnivore club, but bears eat plenty of plants.
Cats don't have any grinding-style molars because we don't need 'em! Hypercarnivores *only* want *meat, meat, meat*!

The most famous teeth in all of catdom are those of the **saber-toothed cats!** Many prehistoric species around the world had these long, dangerous canine teeth.
The three species of **Smilodon** deserve a special mention.
These teeth are bonkers, right? They look like they'd get stuck shut!
Smilodon gracilis
Smilodon fatalis
Smilodon populator
What, did *Smilodon* hunt with their mouths closed?
poink!
I don't think I could open my mouth wide enough to **eat** if my teeth looked like that!

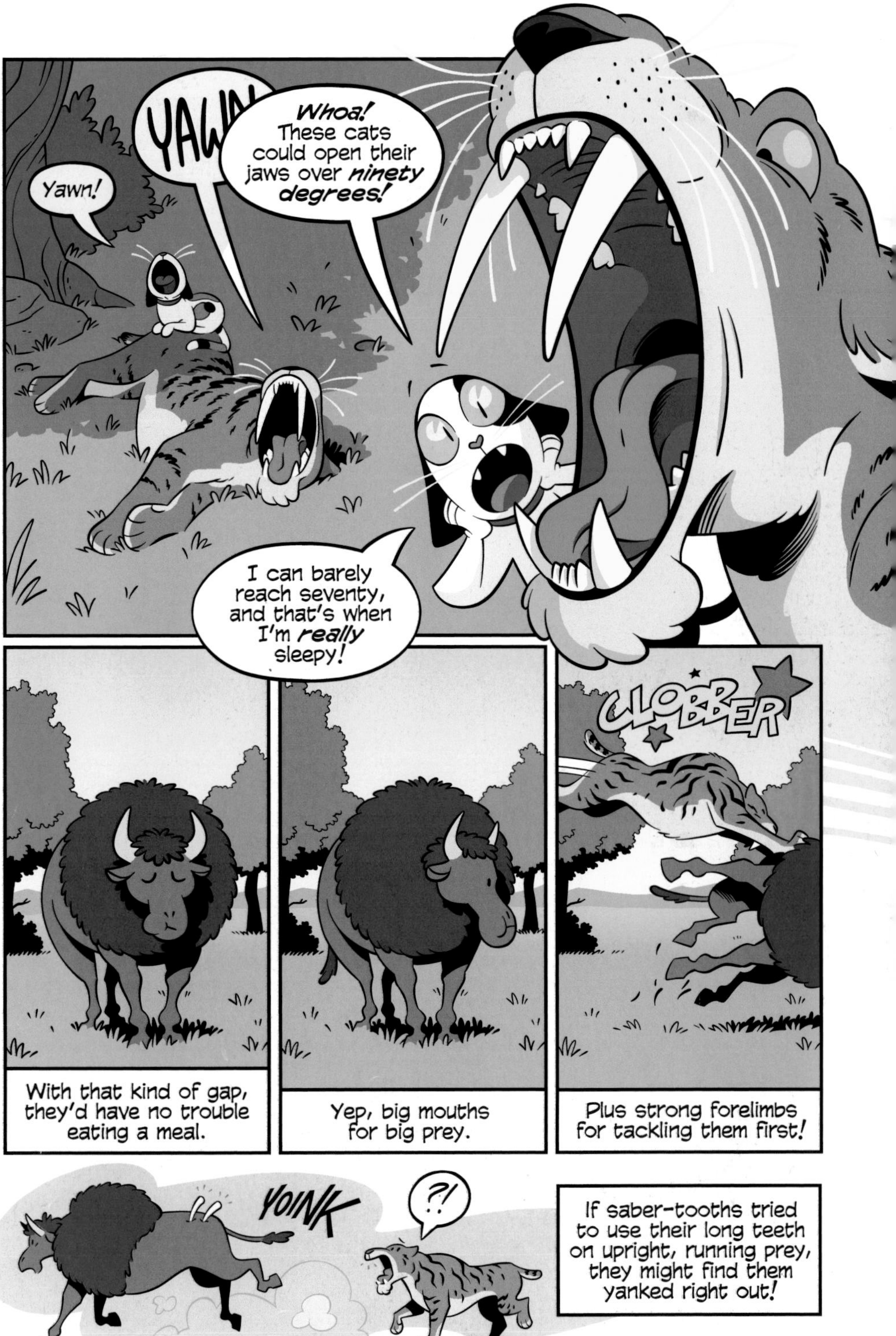

Yawn!

YAWN

Whoa! These cats could open their jaws over ***ninety degrees!***

I can barely reach seventy, and that's when I'm ***really*** sleepy!

CLOBBER

With that kind of gap, they'd have no trouble eating a meal.

Yep, big mouths for big prey.

Plus strong forelimbs for tackling them first!

If saber-tooths tried to use their long teeth on upright, running prey, they might find them yanked right out!

That restraining tackle is still an essential part of the cat family's hunting method today!
In fact, we're the only carnivores who *grab* prey with our forepaws before chomping.
Not that I was having any luck with that.
PAF
MROWR!
PAF
chirp
MROWR!
PAF
chirp
MROWR!
PAF
chirp
MROWR!
PAF
chirp
MROWR.
PLAT
You can see why I was frustrated.

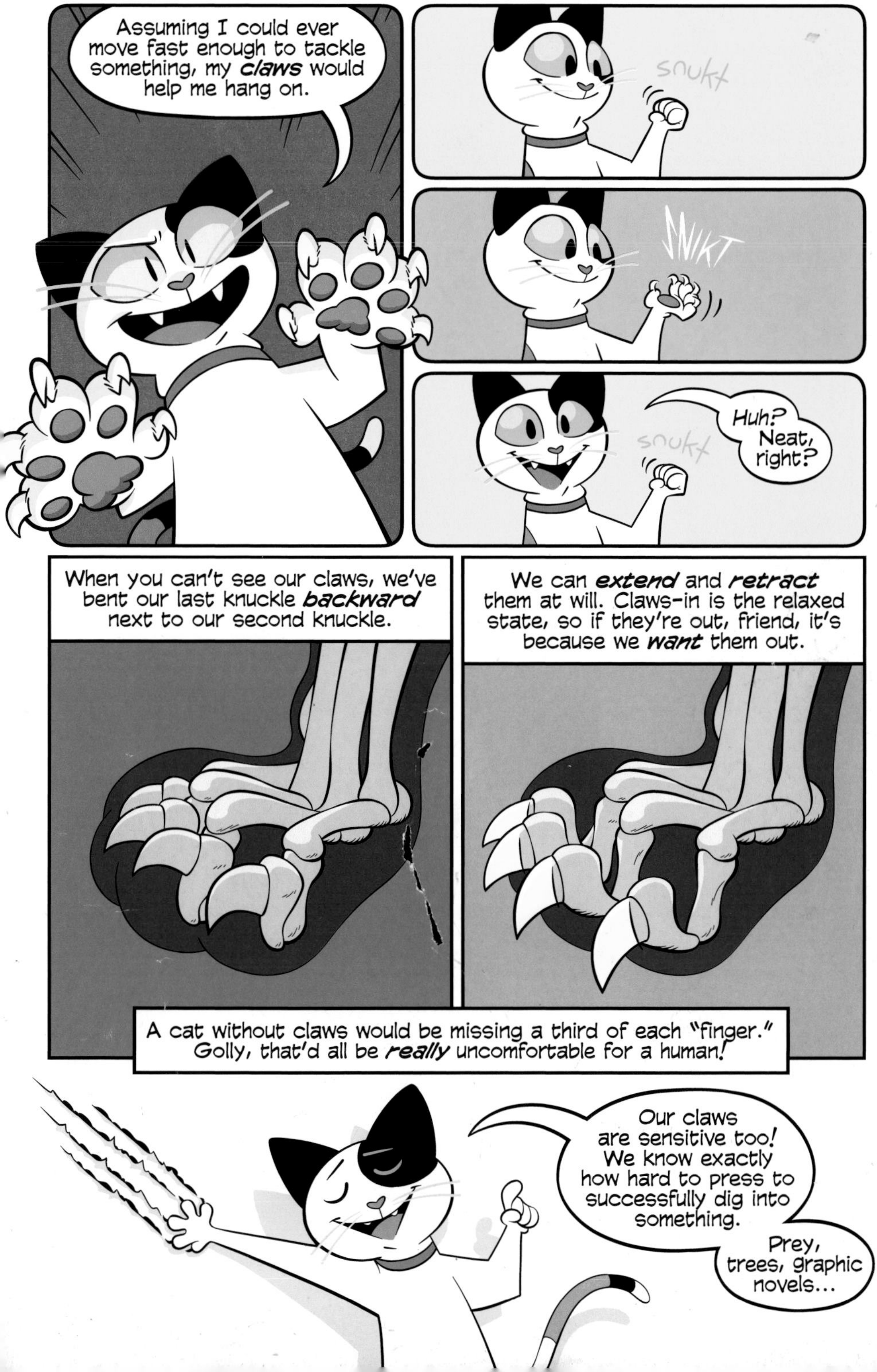

Assuming I could ever move fast enough to tackle something, my ***claws*** would help me hang on.

snukt

SNIKT

snukt

Huh? Neat, right?

When you can't see our claws, we've bent our last knuckle ***backward*** next to our second knuckle.

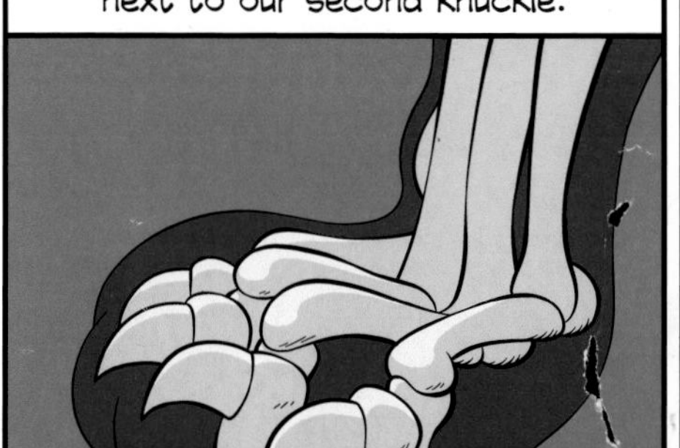

We can ***extend*** and ***retract*** them at will. Claws-in is the relaxed state, so if they're out, friend, it's because we ***want*** them out.

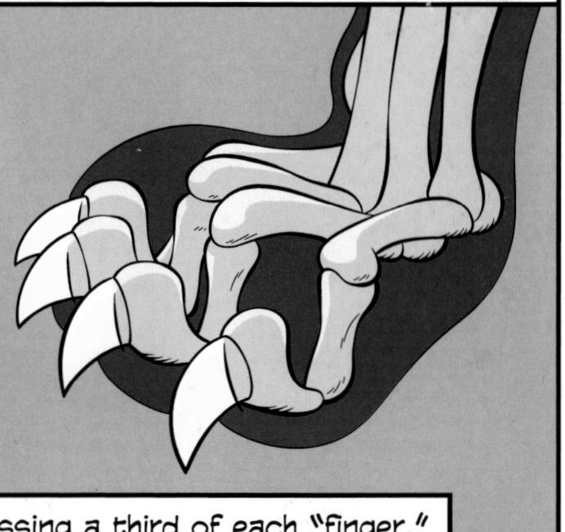

A cat without claws would be missing a third of each "finger." Golly, that'd all be ***really*** uncomfortable for a human!

Our claws are sensitive too! We know exactly how hard to press to successfully dig into something.
Prey, trees, graphic novels...

Teeth, claws...these tools are shared by all cats, from the biggest to the smallest. A 500-pound tiger, a 5-pound kodkod, and all of us in between use them to hunt in the *same way.*
Stalk...
...ambush...
...tackle...
...eat!

A notable exception to that pattern is—

WHOA!

the cheetah!

Cheetahs like to ***chase!*** They're the fastest land animal, with top speeds of ***97 kph (60 mph)*** while clearing over ***6 meters (20 feet) per stride!***

They can only keep this up for a short while, but ***any*** time is longer than these antelope would like.

Ack! Everybody panic!

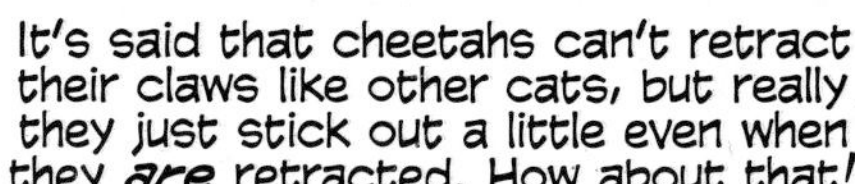

It's said that cheetahs can't retract their claws like other cats, but really they just stick out a little even when they ***are*** retracted. How about that!

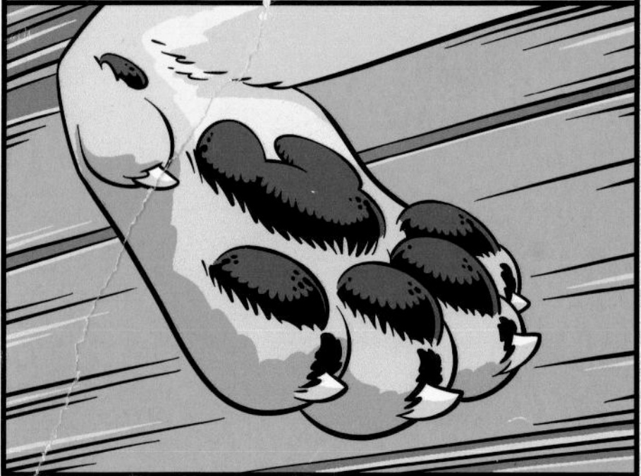

You can't shake me!

During a sprint, fully extended claws act like ***cleats*** to give the cheetah extra traction.

Now, look away if you must, but hunting is an ***essential part*** of cats' nature.

All that meat has to come from somewhere.

STOP! There are kids reading this!

Most cats take down relatively large prey with a tackle and then a suffocating bite from below.
Sleep... ***sleeeep,*** my supper...

With smaller prey, though, cats follow up their tackle with a surgically precise killing bite. A single sensitive canine tooth wedged between pieces of their prey's backbone, and...***POP!***

It's over before they know it.

Though, *uh*, that doesn't apply if your prey doesn't have a backbone.
chew chew

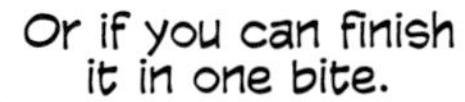
Or if you can finish it in one bite.

GULP

grrrowwl
Sigh.

I can't say I'd found myself in the most ***generous*** hunting grounds.

399
A good environment has a hundred pounds of prey for each pound of cat, so one or two crickets don't do much to tip the scales.

I'd say this was a tough spot to find myself in, but other cats make do in *far* more extreme circumstances.
STILL *SO* DOWN-TO-EARTH

Take the ***sand cat,*** for instance.

These li'l fellas dig burrows to shelter from desert temperatures, both hot—***over 37°C (100°F)*** on a summer day...

...and cold—below ***-17°C (0°F)*** on a winter night.

OW, OW! HOT-HOT-HOT HOT!

The sand itself can get to be ***79°C (175°F)***, so these cats grow protective hair mats over their paw pads. Jealous!

When water is hard to come by (*and it's always hard to come by*), sand cats aren't worried. They get all the moisture they need from their prey!

At the other extreme, the *snow leopard* lives high atop frigid mountains.
Here, over a mile above sea level, the air is thin and cold, but this leopard's overlarge nasal cavities help them stay oxygenated.
INHAA
AAH
ALE
To keep warm, their hair grows much longer than most wild cats', giving them ***extra insulation.***
CRUN
Like the sand cats' feet, theirs are something special.
Snow leopards' ***huge*** paws work like snowshoes on deep soft powder.
CRUN
CRUN

Even species of cats that are closely related to one another, like the different kinds of *lynx*, can adapt to a wide range of environments.

Look how fluffy the ***Canada lynx*** has to be to stay warm in the snow!

Meanwhile, the ***red lynx,*** or bobcat, gets away with a shorter coat.

Red lynx have paws like mine...

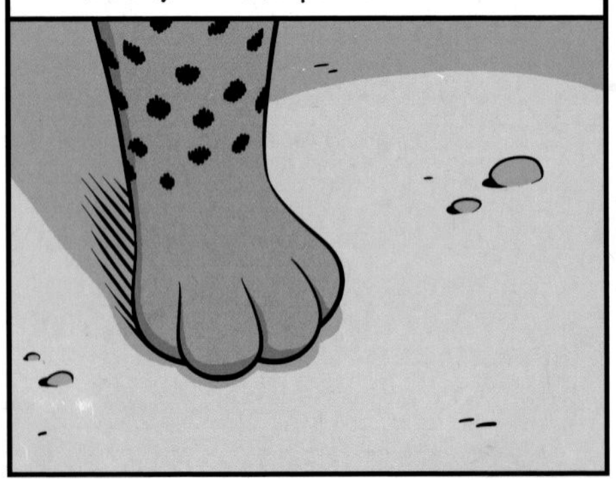

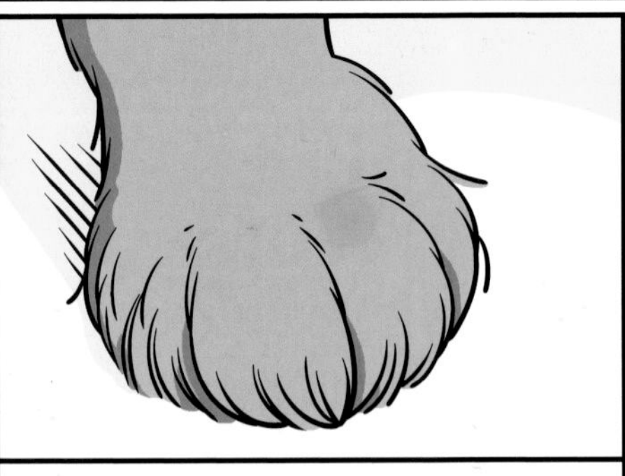

...and Canada lynx have big snowshoe paws like the snow leopard.

Lynx are found all over, and each species looks different to suit ***their*** part of the world!

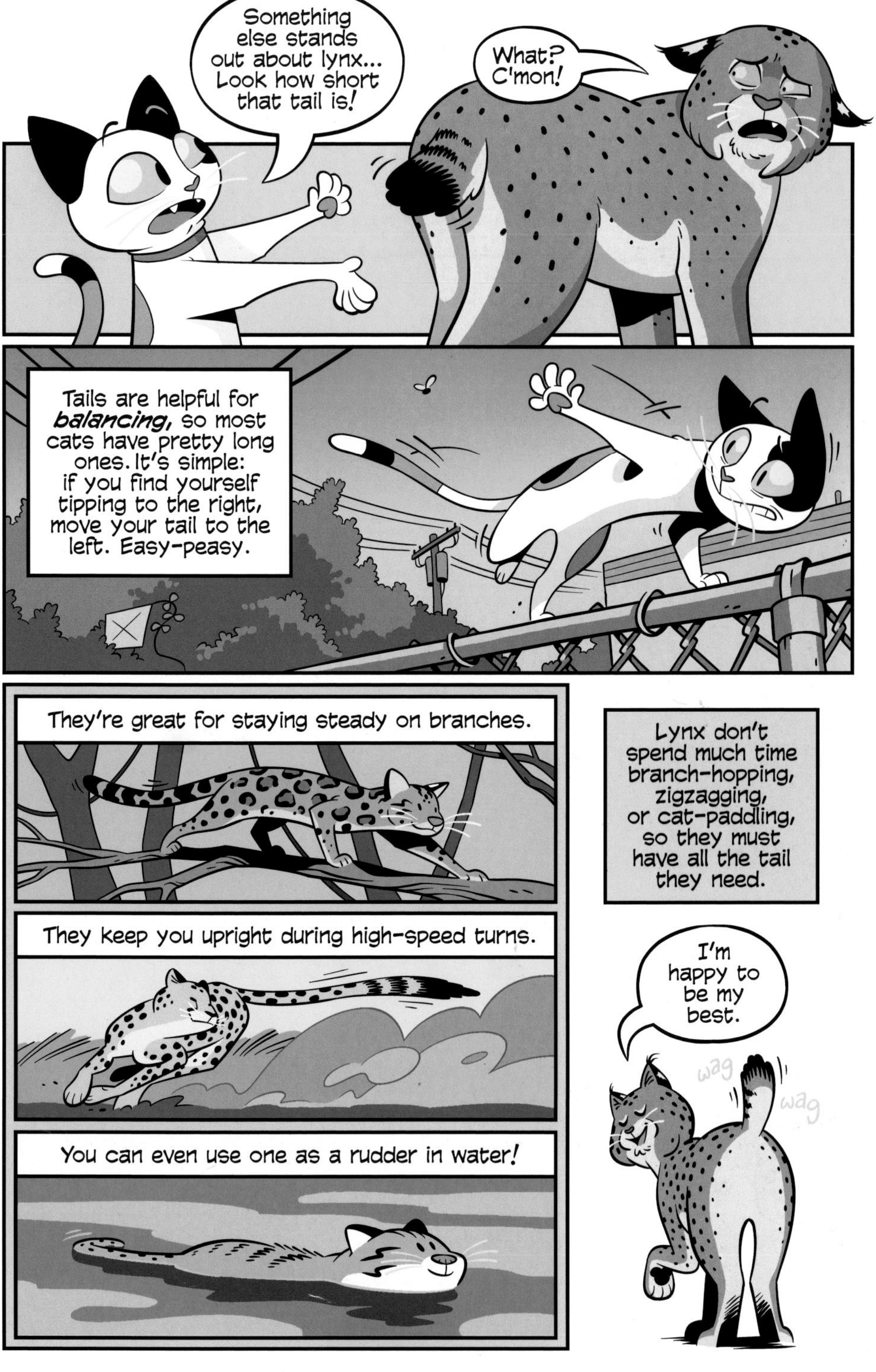

Something else stands out about lynx... Look how short that tail is!

What? C'mon!

Tails are helpful for ***balancing***, so most cats have pretty long ones. It's simple: if you find yourself tipping to the right, move your tail to the left. Easy-peasy.

They're great for staying steady on branches.
They keep you upright during high-speed turns.
You can even use one as a rudder in water!

Lynx don't spend much time branch-hopping, zigzagging, or cat-paddling, so they must have all the tail they need.

I'm happy to be my best.
wag
wag

What's that look for?
"Afraid of water"? No, no, no.
H_2SO WHAT?

Plenty of cats love water! There's prey there, right? ***Fishing cats*** are the best swimmers of us all, and lots of their favorite foods live in the water. Fish, frogs, ducks...it's a buffet!

Sometimes it's nice just to cool off for a little while! Nothing wrong with that.

Wet fur is cold fur, though, which might explain why some cats can be reluctant to get soaked.

We domestic cats like it warmer than average, so I'll stick to my sunbeam.

And, well, my kittenhood experiences with water weren't fun bath times.

KA-RACK

You don't forget that.

Hungry, tired, and now ***wet?*** I know how this story goes, and I ***still*** feel sorry for me.

But I was talking about tails, the most *spectacular* of which might belong to...
...the *clouded leopard!* Theirs is as long as the whole rest of the cat!
Half body, half tail, all pretty!
It shouldn't be surprising that they have such great tails since they spend so much time in trees.
They're acrobatic enough to pull off stunts no other cat can, like climbing down a trunk headfirst!
Or moving along branches upside down!
Or even hanging by their hind paws!
Yet with all that talent, the clouded leopard is named for their *uniquely patterned coat.*

A cat's coat pattern is a good clue as to where they call home.
Patterned coats are most common on cats who live in places with dense bushes or closed-canopy forests.

Solid coats are usually found on cats who live in open environments like grasslands or mountains.

There are some exceptions, of course.
Who, me?

What do you suppose is the point of different coats? Are they just for looking pretty?

Well?
Surprise!
Our coats are background-matching *camouflage!*
MEOW!
Now it makes sense. Solid coats blend in well with rocks and grass.
Spotted coats are almost invisible when sunlight shines through layers of leaves.
What about the *tiger's* unique coat?
The vertical stripes break up their shape, so prey can hardly tell where they begin, where they end, or if anyone is there at all. That goes double if the prey has got poor color vision!

Melanistic individuals have very dark coats and can be found in many cat species.

Cats with this coloration might find it useful for sneaking around habitats with variety. Whether you're around trees, brush, or rocks, there'll always be *shadows* to hide in.

Melanism is common in jaguars and leopards. You might hear either of these two cats called a *black panther!*
What about cougars?!

No melanistic cougar (or puma, or mountain lion, or any other name this cat goes by) has ever been proven to exist, yet there have been ***thousands*** of "black panther" sightings in the United States.

The only big cat native to these parts is the cougar, so how can this be?

Uh-huh, stuck in your tree.
Uh-huh, ***how*** big?

What people report as an American black panther is likely a big domestic cat...

...a shadowy-but-ordinary brown cougar...

...or all in the viewer's imagination.

If black cougars ***do*** exist, they're fantastically, extraordinarily, ***unbelievably*** good at hiding!
I know ***I'd*** never miss one, though!

But poor li'l me.
I didn't have spots to blend in with bushes.
Dangit!
I wasn't dusty-colored to hide in the open.
Dangit!
I sure stuck out in the shadows.
DANGIT!
MREEOOOW!!
No, there wasn't much use for calico camouflage, and my belly wasn't shy about telling me so.
grrrowl
I know, belly. I know.

I couldn't believe it either! You know how far people would go to stroke this fur today?
Ridiculous!
FAMOUSLY PETTABLE KITTY SADLY ONCE UNPET

But still, a cat is more than her coat. More than her teeth and claws too.

I wasn't ready to give up!
I'm not ready to give up!

It was time for me to use every tool I had to get the meal that would keep me going!

You can't defeat me, world!

So I took a deep breath and focused on all my ***natural catty abilities!***

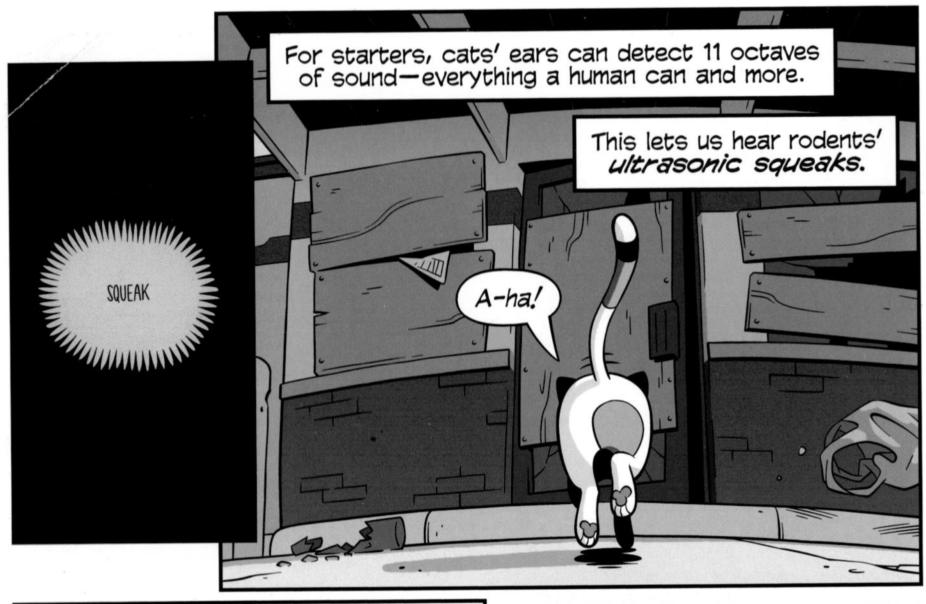

SQUEAK
For starters, cats' ears can detect 11 octaves of sound—everything a human can and more.
This lets us hear rodents' *ultrasonic squeaks.*
A-ha!

We can tell what direction a sound is coming from by which ear it hits first.

Our ears turn independently to tune in more precisely.

SQUEAK
A little that way!

SQUEAK
Up there!

We can even tell how low or high the source is thanks to special ridges in our ears. That's right, ***3-D hearing.***

Servals in particular are known for their exceptional hearing.
SHH!
It's no surprise with ears like ***that.***

These lanky cats have been known to sit for long times just waiting and ***listening*** for rodents among the tall grass.

Once they hear their prey, servals use their long legs to pounce on them.

At only 60 cm (2 ft) tall, they can jump almost ***3 m (10 ft) straight up!***

SQUEAKA SQUEAK

They're so dependent on sound that they won't bother hunting on windy days, when rustling grass hides their prey's squeaks.

I could use a nap anyway.

Cats' eyes are incredibly sensitive to *light*, which makes us extra good at hunting in near *dark*.
Where are you at, you future snack...?

Vision works when light shines onto cells in the back of the eye called *photoreceptors*.

The brightness-sensing photoreceptors are *rods*, and cats have *lots* of them. So many that even the tiniest sliver of light is likely to be detected.

And if there's *still* not enough light for our rods, well, we'll multiply it.

The tapetum lucidum sits behind our photoreceptors and reflects light past them a second time.

It's what makes our eyes seem to *glow* in the dark!

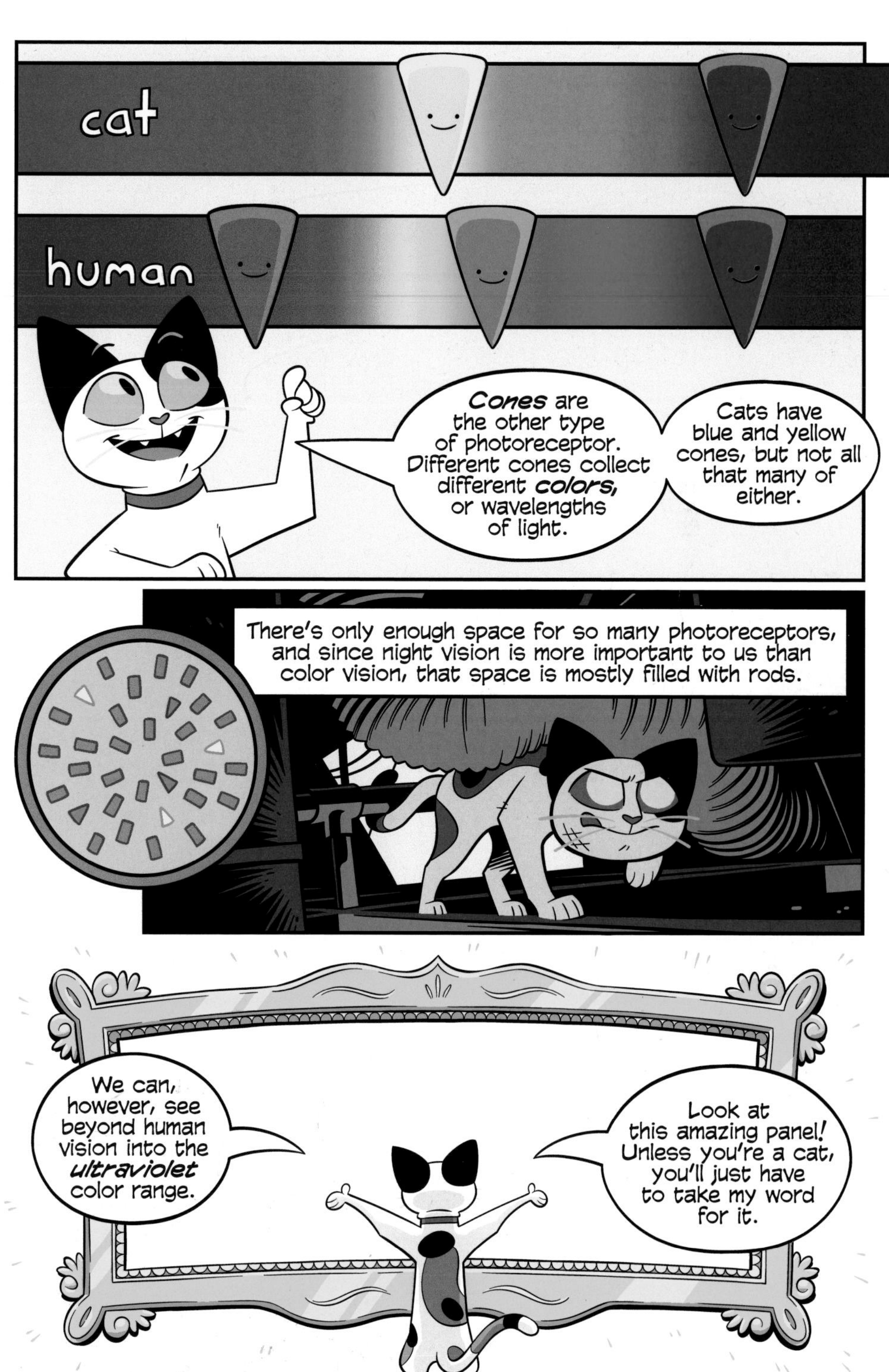
cat
human
Cones are the other type of photoreceptor. Different cones collect different *colors*, or wavelengths of light.
Cats have blue and yellow cones, but not all that many of either.
There's only enough space for so many photoreceptors, and since night vision is more important to us than color vision, that space is mostly filled with rods.
We can, however, see beyond human vision into the *ultraviolet* color range.
Look at this amazing panel! Unless you're a cat, you'll just have to take my word for it.

CLANG!

We only want to let in so much light at a time or all those sensitive rods will get overloaded.
By ***contracting*** and ***dilating*** our pupils, we can control just that.
Ack, bright!

When it's bright, we want to contract our pupils to a ***small*** size so we don't get more light than we can handle.

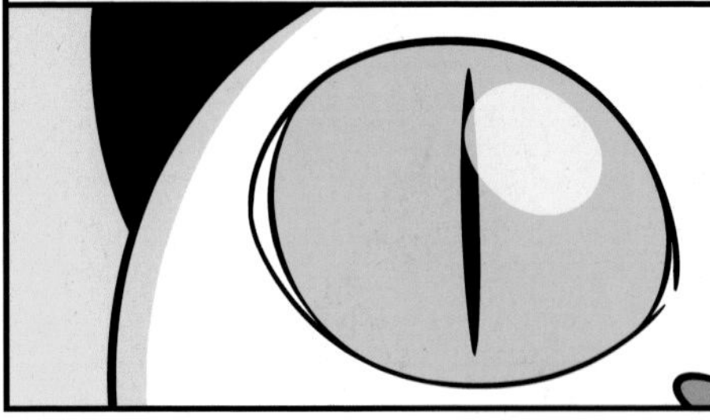

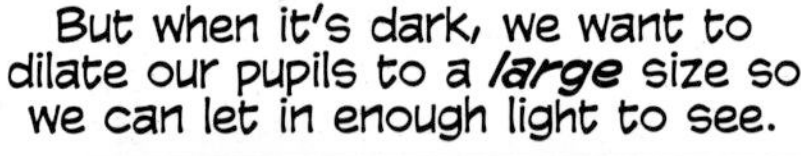

But when it's dark, we want to dilate our pupils to a ***large*** size so we can let in enough light to see.

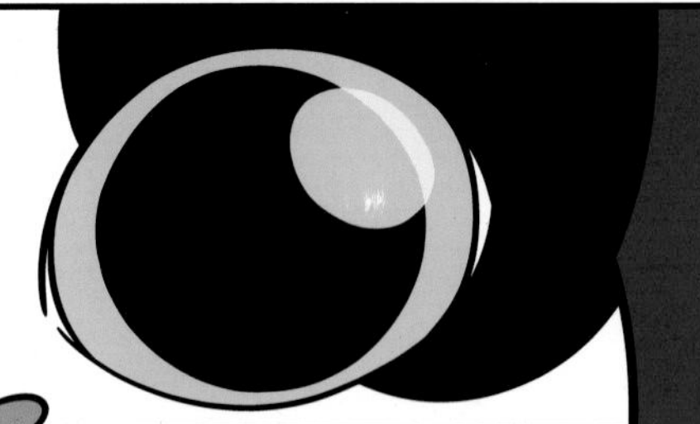

Humans can dilate their pupils to let in 15 times as much light in the dark.

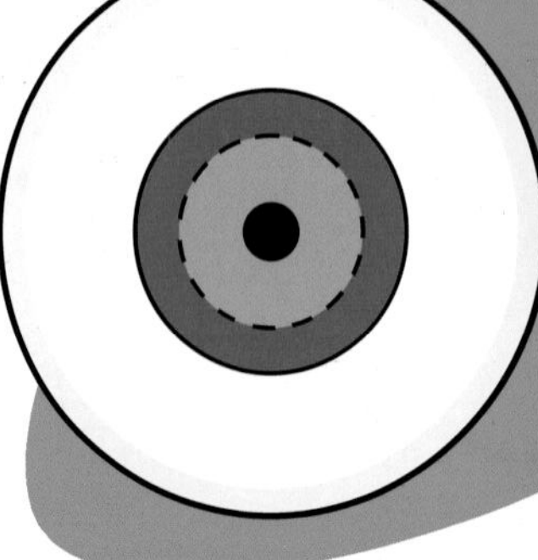

Not bad, but domestic cats like me can dilate our pupils enough to let in ***135 times*** as much light.

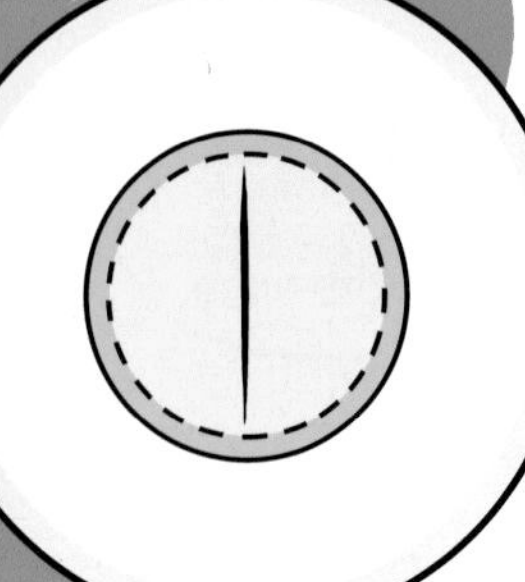

No wonder you never see a cat with a flashlight!

Slit pupils are considered something of a trade-mark cat feature, but not all cats have them.
In fact, the larger a cat species is, the more likely they are to have ***round pupils.*** Why might that be?

Slit pupils produce an exaggerated blurring effect on horizontal objects.

For a low-down eyeline, this means it's very easy to judge how ***far away*** something on the ground is.

The effect is less pronounced the higher up your eyeline is, so slit eyes are less useful to tall cats.

Most cats are ***ambushers.*** It's important to land that first pounce, since we might not get a second!

Even without the slit pupil advantage, all cats can judge distance thanks to our *binocular vision.*

Binocular vision is common to all animals with front-facing eyes, even humans. It works by combining the view from the left eye...

...with the view from the right eye...

...to make a *3-D image. Hi-ya!*

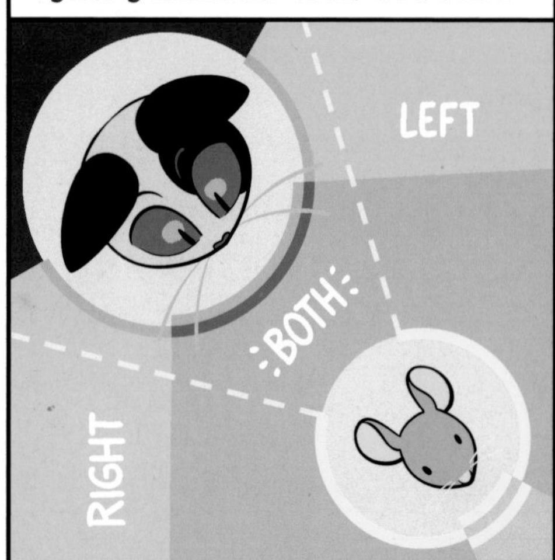
The more your two eyes' fields of vision overlap, the better *depth perception* you'll have. Prey animals tend to have eyes on the sides of their heads to expand their total field of vision side to side and hopefully avoid getting ambushed. *Hmm,* we'll see...
LEFT
BOTH
RIGHT

SQUEAK
Combined with our stereo hearing, cats are acutely aware of what's where in their surroundings.

Remember, we want to get this pounce just right.

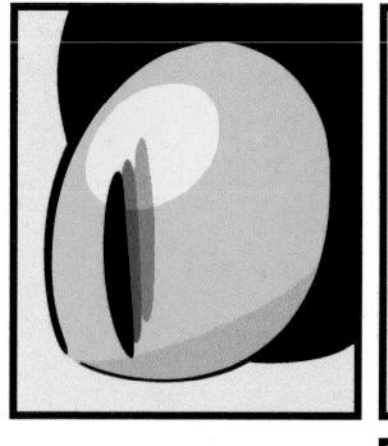

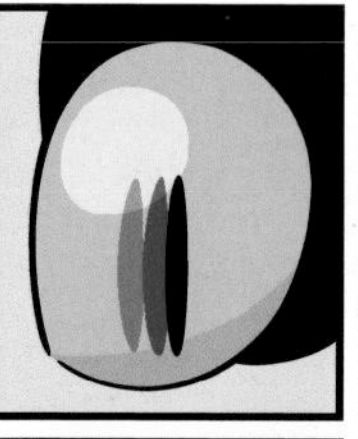

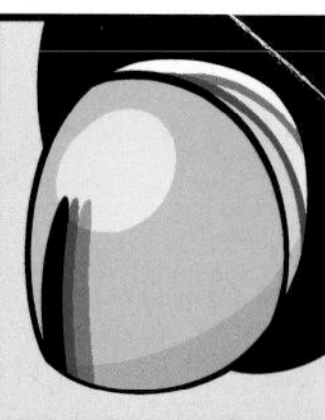

To follow quick prey, cat eyes make ***saccades,*** rapid movements that prevent motion blur.

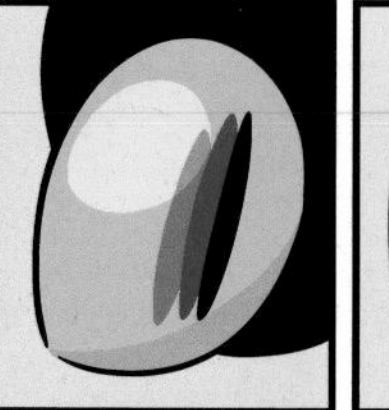

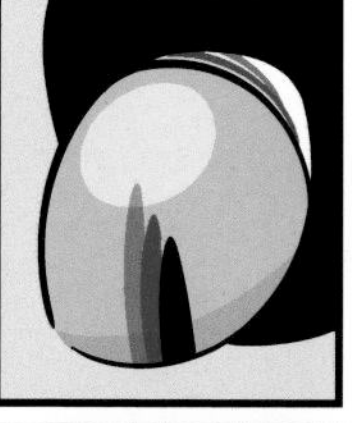

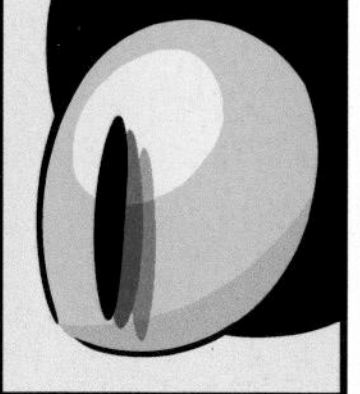

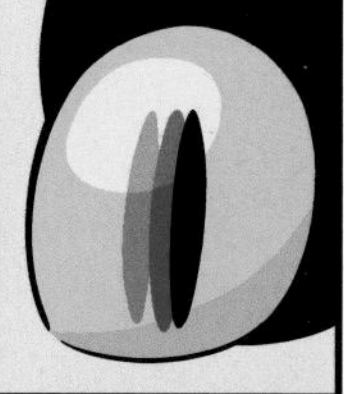

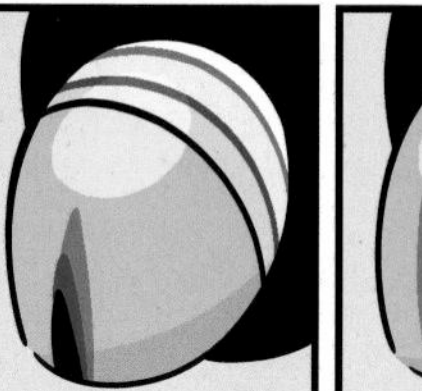

In concert with a brain wired to prioritize motion...

...this mouse isn't gonna get away!

Slow this down! ***Way down!***

In this final fraction of a second, my most important tool isn't my beautiful eyes or my adorable ears.
It's not my fearsome claws or my fearsome teeth.
It's my secret weapon: my ***whiskers.***

Whiskers are stiff, *very* sensitive hairs on a cat's snout, brow, and wrists.
How sensitive? They can detect teeny-tiny changes in the *air* around them.

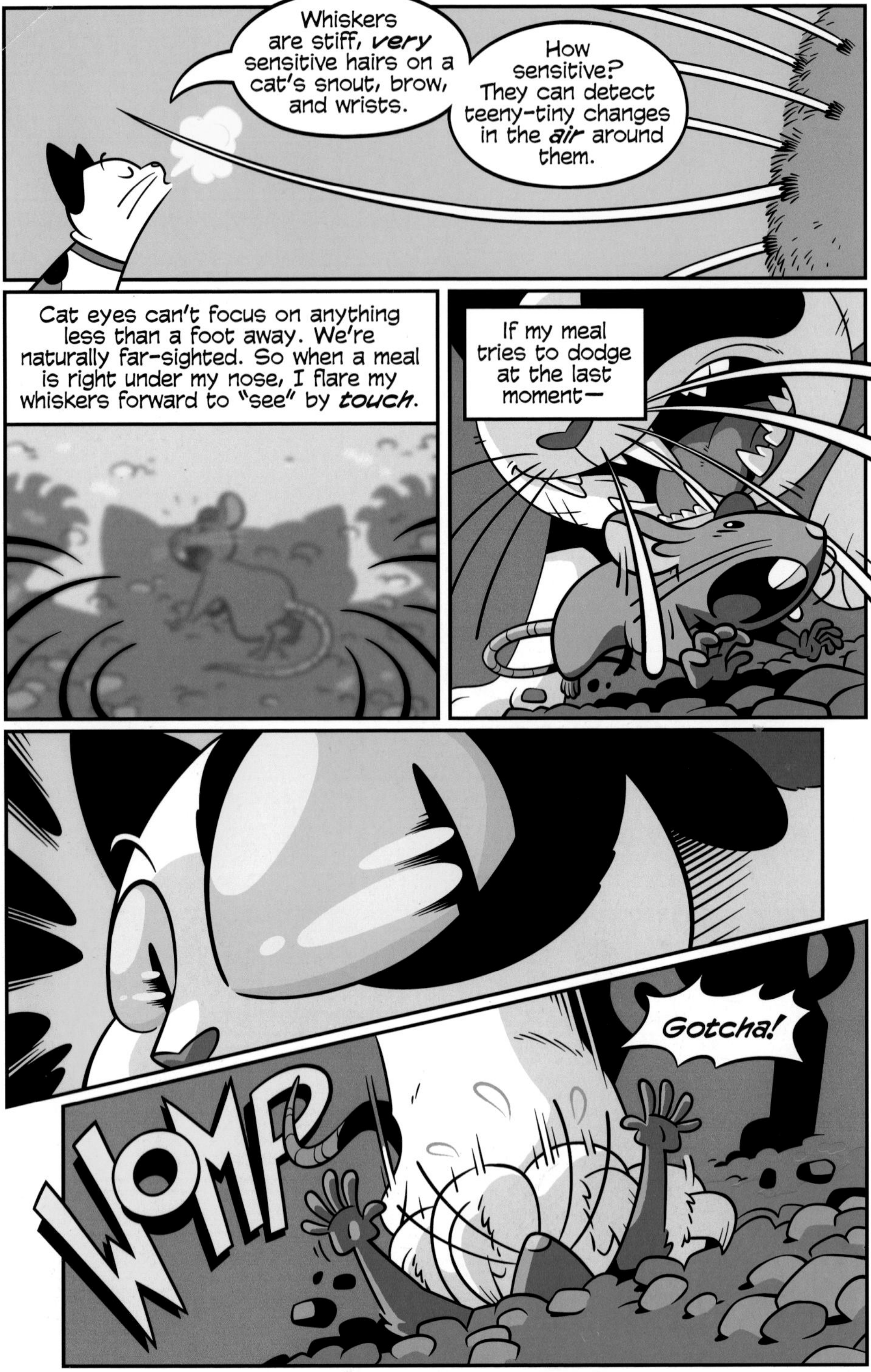

Cat eyes can't focus on anything less than a foot away. We're naturally far-sighted. So when a meal is right under my nose, I flare my whiskers forward to "see" by *touch*.

If my meal tries to dodge at the last moment—

WOMP

Gotcha!

GULP

GULP

Delicious.
Like the rest of our senses, cats' sense of *taste* is attuned to—what else?—meat!
GETTING A TASTE FOR IT

We get all the nutrition we need from it, so while we're pros at distinguishing between meats, our palette is pretty insensitive to anything else.

sniff
sniff
sniff

What is *this?!*
WHAAAAAAAAAAAAAAAA

SWAT
I know I'm a *picky eater*, and I don't care. If I didn't hunt a meal myself, you're darn right I'll side-eye it!

In the wild, cats aren't above scavenging. We're very good at finding a balanced diet, and picky eating helps us make sure that any one meal won't be harmful to us in the long run.

If something makes us sick once? ***Never again!*** Strike it from the list and throw 'em both in the trash.

It's true that cats, especially kittens, will eat grass now and then. You could call this "health food" because it helps us clear harmful parasites from our guts.

URP!

What about that *other* famous plant we love?

Catnip!
Hoo boy! Just smelling that stuff turns on parts of our brain like ZIP!
It only works on cats, so other animals might look at us funny when—KERPOW! We can't control ourselves!
Catnip affects a little more than half of domestic cats, as well as a handful of wild cats, and it makes us go bonkers.
Weirdly, catnip doesn't have any influence on cats who are native to the same parts of the world as it is. Scientists chalk the whole thing up to a strange coincidence.
Other plants such as honeysuckle, valerian root, and especially silver vine have the same effects on many cats, including some who don't respond to catnip.
Dopes.
Ready for a wild theory? Scientists think the skin secretions of stressed-out mice can cause a similar response, giving them one more chance to escape!
Uh, what was I doing again?

I promise I was as picky back then as I am now.

A little hungrier, but still picky.
Still?!
GRRROWL!

SQUEAK
SQUEAK
Ah!

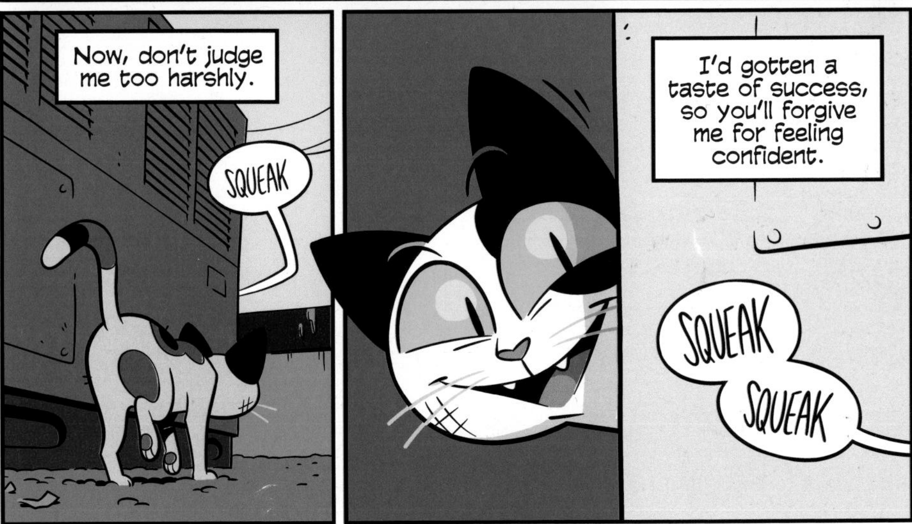
Now, don't judge me too harshly.
SQUEAK
I'd gotten a taste of success, so you'll forgive me for feeling confident.
SQUEAK
SQUEAK

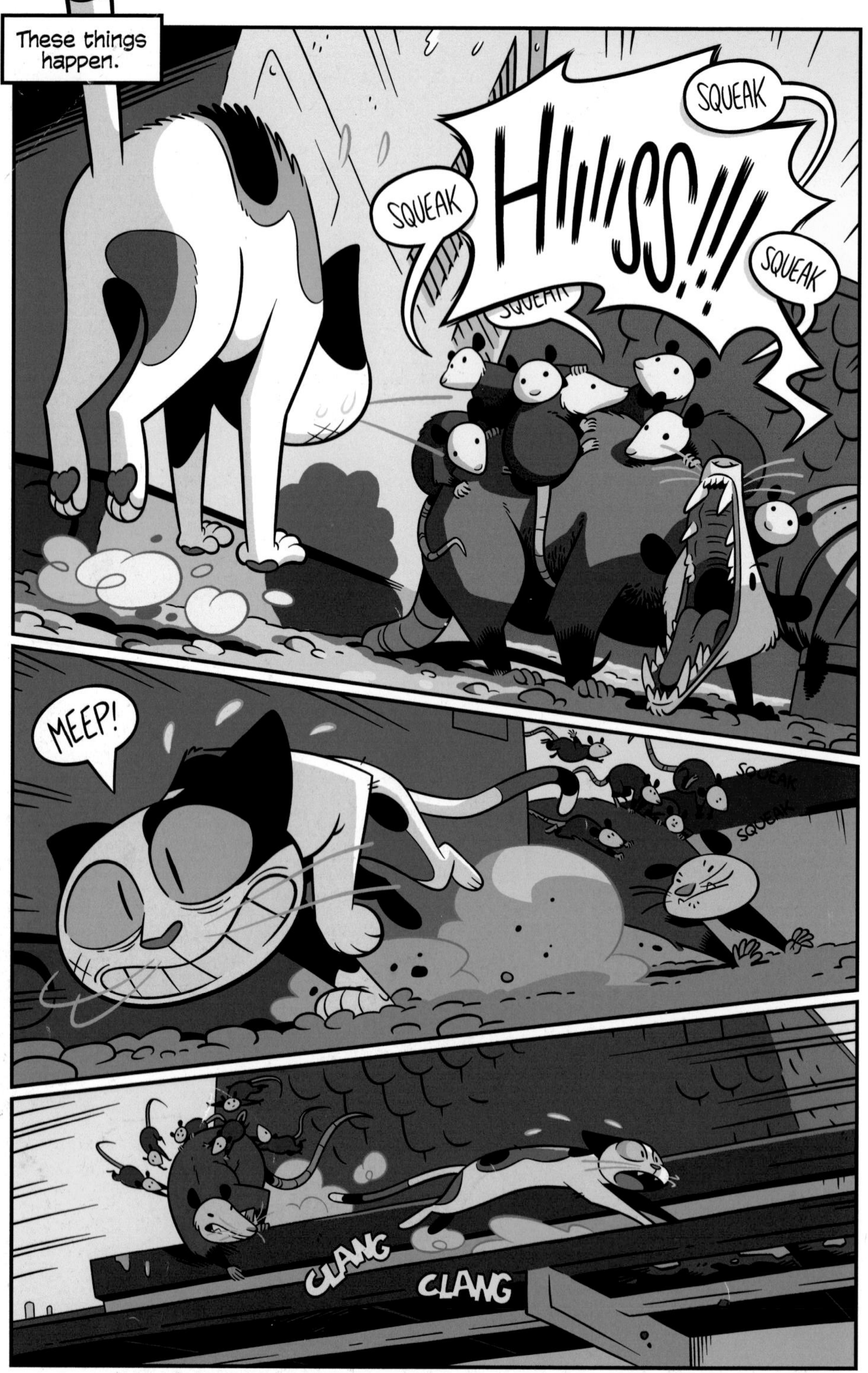

These things happen.
SQUEAK
HISS!!!
SQUEAK
SQUEAK
MEEP!
SQUEAK
SQUEAK
CLANG
CLANG

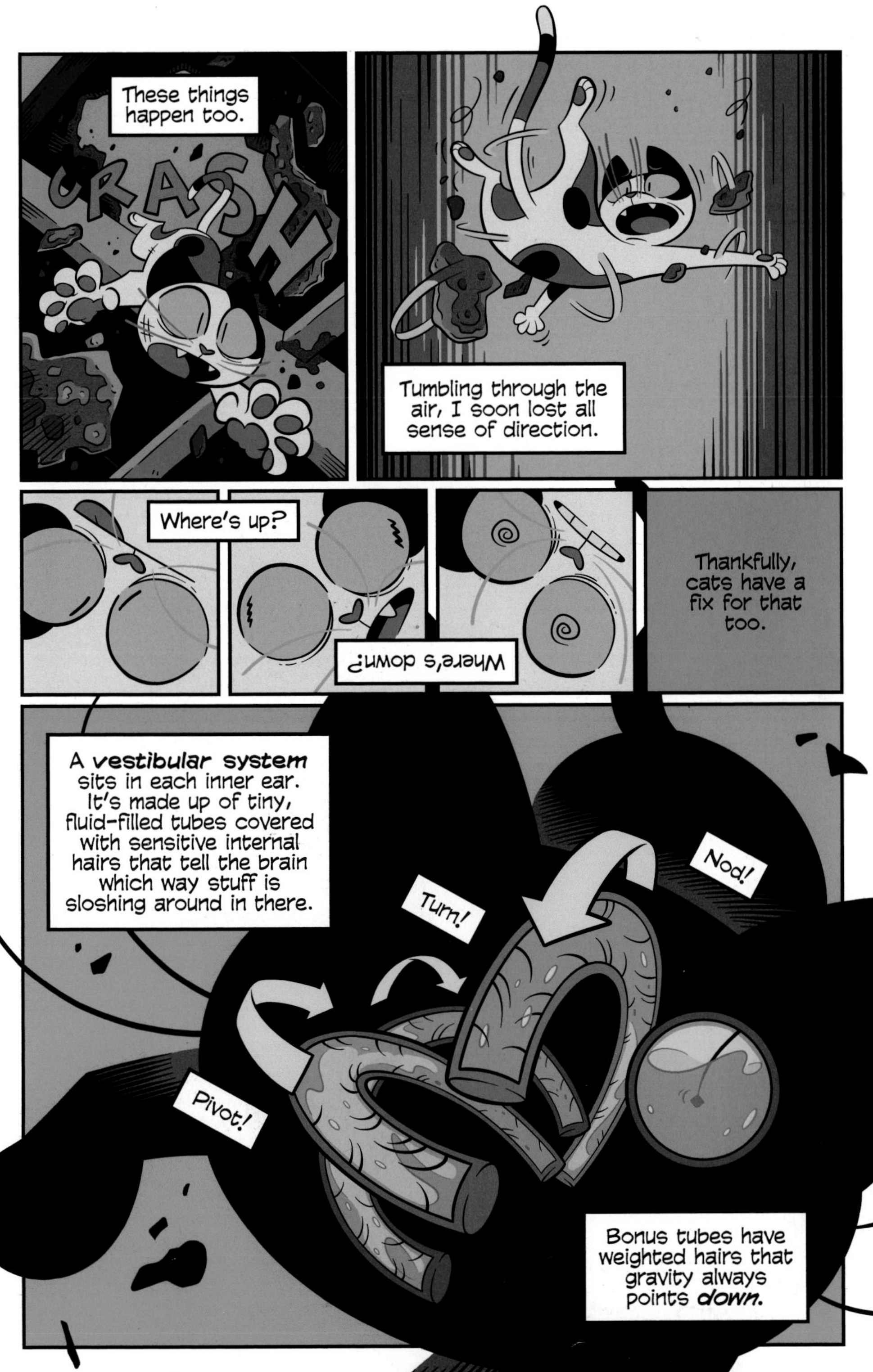
These things happen too.
CRASH
Tumbling through the air, I soon lost all sense of direction.
Where's up?
Where's down?
Thankfully, cats have a fix for that too.
A ***vestibular system*** sits in each inner ear. It's made up of tiny, fluid-filled tubes covered with sensitive internal hairs that tell the brain which way stuff is sloshing around in there.
Turn!
Nod!
Pivot!
Bonus tubes have weighted hairs that gravity always points ***down.***

Within a tenth of a second, I knew which way was up again. Now the cat righting reflex took over to make sure I landed on my feet.
First, I turned my head to find the ground. Then I tucked my front legs in and twisted my flexible spine.
I kept my back legs extended so I could use air resistance like I was pushing off of nothing.
Yeah, it's pretty slick.

Next, I extended my front legs to slow my roll and tucked my back legs to *repeat* my front twist.
With all of me pointing the right way now, my only problem was that I was *still falling*.
Whoosh! Parachute pose!
Everything you need to know is in the name. This move limits speed enough to give a cat a fighting chance of at least *surviving* a fall of many stories.
Sufficiently slowed, I just had to stretch out all four legs, arch my back to cushion my landing, and...

10
10
10
HISS! HISS!
The righting reflex is a real lifesaver, but it takes some time to pull off. *Short* falls can be even more dangerous than long ones, and *any* fall could result in injury.
NYAH!
Lucky me!

A sorely needed meal caught here.

Danger narrowly avoided there.

On good days, I found safe places to rest. On bad days, I slept with one eye open.

One particularly tough time left me so long without a catch that I could barely carry my own li'l body.

sniff
SNIFF SNIFF
Wh-what's that?

If a cat doesn't see or hear anything, there's still a chance they can *smell* something. And smell something I did!
We've got such *boopable* li'l noses that folks underestimate how keen our sense of smell is.

Cats have ***hundreds*** of different scent receptor types, each of which can detect varying intensities of its target odor. In combination, these could potentially distinguish between ***billions*** of scents, more than a cat will ever smell! Not bad!

I followed my nose until some *scratches* on a post caught my eye.
Ah!
Oh, that? That goofy face I'm pulling is called a *flehmen*.
It means I'm using my special secret second scent-detector, my *vomeronasal organ*.
This li'l thing sits up in the attic of my mouth, you know, above the roof, and it detects certain chemicals called *pheromones*.
Pheremones are unique to each species. They communicate social information like identity, health, and if a cat is looking to have kittens. They're like a *scent profile card*.
AGE
HEALTH
ATTRACTIVENESS
Huh, so that's who you are.

Those scratches were there to draw extra attention to a smell. Someone really wanted to be noticed!

But *where* are you?

Scent marks are a form of delayed communication, since smells can stick around long after the cat who left them has moved on.

Maybe they know where *food* is!

Wild cats don't often meet face-to-face, and scent marks can help them *keep* avoiding each other. First meetings don't always go well, you know.

Ulp! Maybe they'll want to *fight* me!

I had no choice but to take that chance.
If I don't get something to eat, it's moot!

A li'l kitties' room.

The feline facilities.

Nature's litter box.

Stink storage.

The ol' dirt outhouse.

Stop me, I've got more.

Middens are often found along travel routes or at territorial boundaries. We have the sense to make them well away from anywhere we eat or sleep. A midden is obviously a safe space to do our business *(that's how it ended up used)*, but after a few days, parasites inevitably start to build up. Then we want nothing to do with it!

Meow?
It was the first time I could remember seeing another cat.

What do I say?
What do I do?

A *tail-up* posture seemed right. It's a clearly friendly sign, like waving hello.

So far, so good!

Ah, *bunting*. Nicer than it looks.

Friendly cats greet each other by enthusiastically rubbing against one another.

It's not a massage, though. We're depositing *our* scent on the *other* cat.

See, we're absolutely *covered* in odor-oozing glands, and we love to spread our smell around.
Tail wrap? Let me introduce myself!
Cheek rub? I was here!
Claw scratch? My territory!
It's not clear if each *gland* has its own unique secretion, but scents *are* personalized to each *cat*.

I couldn't believe that was all!
Hey, wait!
Because of our solo lifestyle, cats aren't known for being naturally communicative with one another.
Wild cats don't collaborate...they *compete.* Let's say a cat found some good eating.
YES!
Hollering about it for everyone to hear will only draw unwanted attention!
Mine!
Hey! *koff* G-give those back! *koff*
What're you gonna do about it, weakling?
And what if they're sick? It's no good to tell a bully you're a pushover. Nope, nature rewards the quite type.

I needed help, though, and I thought this other cat must know something about surviving here.
Slow down!
There's only one wild cat who *is* the social type, and they're a big one...the *lion*. Lions form prides that can include more than two dozen individuals, mostly lionesses and kittens.
Females do most of the hunting together as well-practiced teams.
ROWR!
GRRR!
With cooperation, they can take down large prey like wildebeests, hippos, and even elephants!
!
ROWR!

Meanwhile, males stay home and rest **20 hours** out of every day.
I'm not sleeping, I'm just resting my eyes.

O-oh! Welcome home!
Let me guess...
Of all the wild cats, lions are the only ones known to give a friendly ***tail up*** like domestic cats.

If any other wild cat species is in a group, it's probably a mom and her kittens. They'll grow up and move away before long, I'm sorry to say.
She hit me!
Did not!
Did so!

Nice, bro.
Yeah, bro.
Love you, bro.
The only other species to form long-term groups is the cheetah, who'll hunt with his brothers even into adulthood.

Was I ever going to have anything like that?

Gosh...maybe this wouldn't be such a bad day afterall.
Vito's
PARKING
ONLY
CAUTION

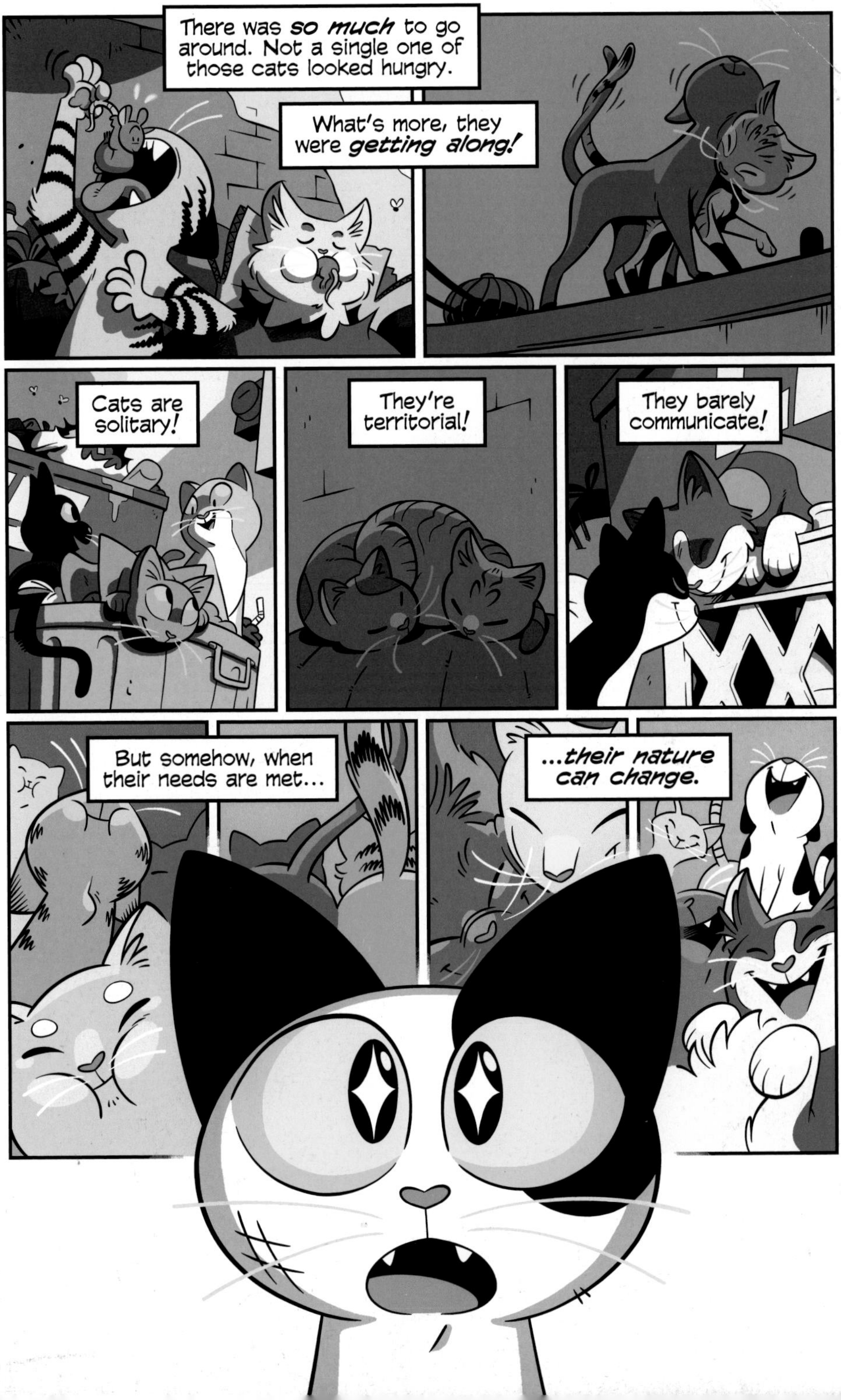

There was *so much* to go around. Not a single one of those cats looked hungry.
What's more, they were ***getting along!***

Cats are solitary!

They're territorial!

They barely communicate!

But somehow, when their needs are met...

...their nature can change.

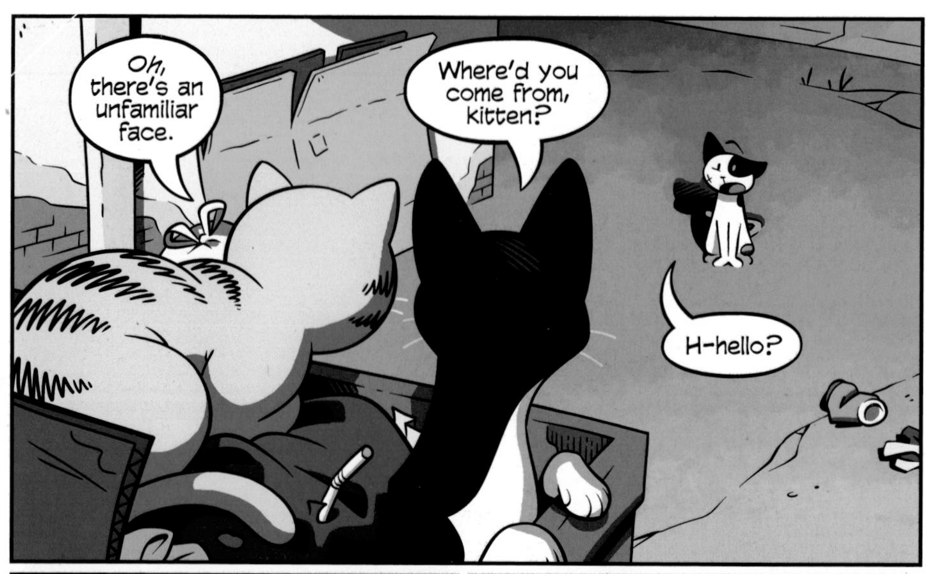

Oh, there's an unfamiliar face.
Where'd you come from, kitten?
H-hello?

I'm *Karen.*
BUNT

Paula.
BUNT

Pleased to meet you.
I'm... new here.
We can see that, kitten.
MEW?!

I sort of just let it happen. In fact, my ***scruff response*** meant there wasn't much I ***could*** do. When a mom carries a kitten by the ***scruff***, the loose skin on the back of their neck, the kitten reflexively goes limp and becomes easy to carry.

Moms don't scruff their kittens to scold them or anything like that. They do it to get the kitten out of danger and back to the nest.

Families move dens a lot in the wild, so it's convenient for kittens to be cooperative.

Here we are! Don't bother unpacking.

I'm talking a lot about moms because they're the ones who stick around. Dads are restless and always wandering off to wider ranges, but moms take care of business at home. With females in charge, cat society is ***matriarchal***.
Cat colonies form around plentiful, concentrated food, and as long as that food remains, one generation's daughters will stick around to become the next generation's moms.

It's a good thing kittens have at least one responsible parent because we're helpless li'l things at first.
Almost there, sweetie.
Newborn kittens can't see or hear, but they *can* smell and, most importantly, sense *warmth*...
Mom!
Mom!
Mom!

...so they can *snuggle up* in groups!

Hello, kitties, we have a new neighbor.
H-hi?
HELLO!

In fact, newborns are so mom-dependent that they'll *freeze* if they don't stay near her, even in warm weather.
Come on, or you'll catch cold!
m-m-mew...

It'll be weeks before they live on anything but milk. They both *need* milk and *knead* for milk. Get it? They use their li'l paws to start dinner flowing.

Newborn kittens can't even *clean* themselves.

No matter their size, grooming is important to cats.

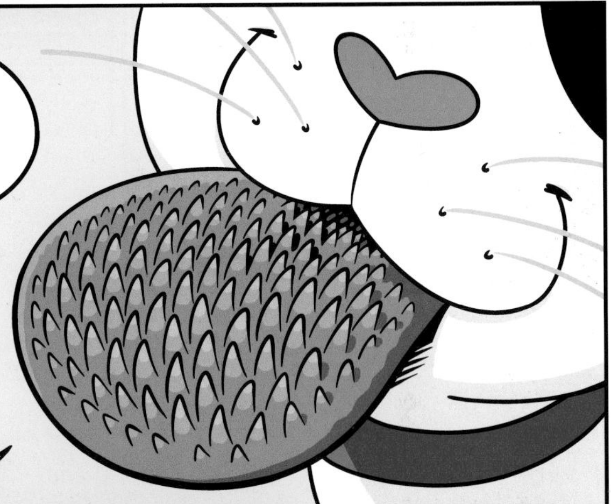
Our tongues have a scratchy sandpaper texture because they're covered in tiny barbs called *papillae.*
Tough enough to let us scrape meat off of bones, these help us comb and detangle our fur, all while raking up loose hairs, old skin cells, dirt, bugs...

All of that gets swallowed, and what doesn't pass through comes out—
ULK!
—'scuse me. Comes out as a—

ULK! ULK!
HAAACK

...*hairball.*
It's a small-cat thing.

Cats only sweat out of our paws, so saliva spread by grooming helps us beat the heat.

And skin oils released by grooming insulate us from the cold and wet.

Let's not forget that a *clean* cat is a *stealthy* cat.
PEE-YOO! Somebody stinks!

We use our paws to get to those hard-to-reach spots...

...and our family and friends add *affection* to hygiene.
Sweet dreams, kitten.

If you could hear those kitties *purr*... It's a quiet request, maybe for mom to simply stay right where she is.
rrrrr puuurrrrrr
PUUUrr
Mom's purring too. The *low vibrations* seem to help her recover from the stress of birthing. It's both an emotional and physical salve.
rrrrrrr
Funnily enough, cats that *purr* can't roar, and cats that *roar* can't purr.
MREOW!
koff
MREOW!
ruh...
ruuuuh...
ruuuhr?

You could even say purring separates so-called "*small*" and "*big*" cats.

RRRROAR

rrrrrrrrrrrrrrrr

PUUUr

At about the same size right in the middle are the biggest purring cat, the puma, and the smallest roaring cat, the leopard.

Shhh! They're trying to sleep up there!

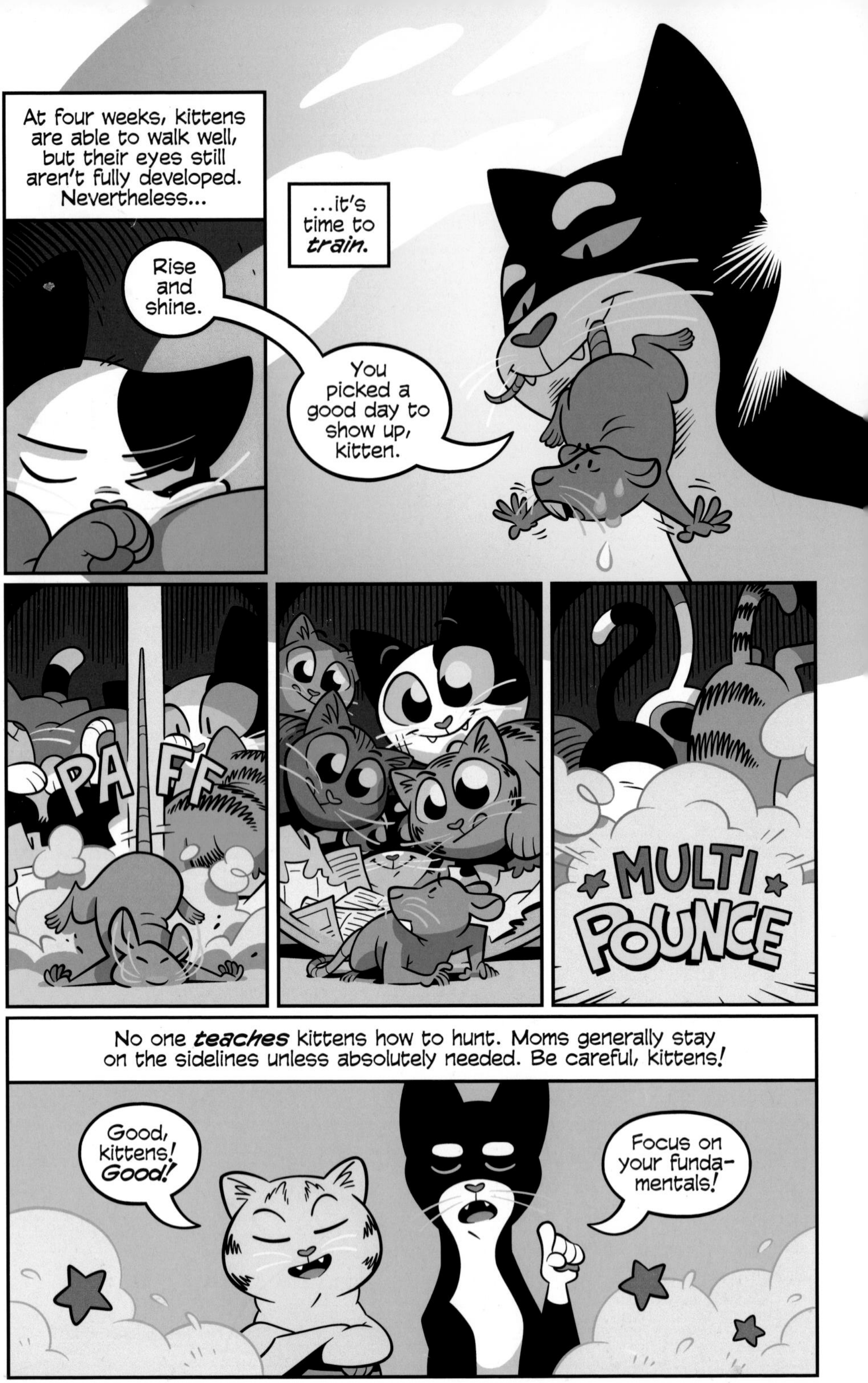

At four weeks, kittens are able to walk well, but their eyes still aren't fully developed. Nevertheless...

...it's time to *train.*

Rise and shine.

You picked a good day to show up, kitten.

PAFF

MULTI POUNCE

No one ***teaches*** kittens how to hunt. Moms generally stay on the sidelines unless absolutely needed. Be careful, kittens!

Good, kittens! ***Good!***
Focus on your fundamentals!

Instead, the behaviors kittens have learned ***playing*** with each other turn into ***hunting*** behaviors when they're applied to prey animals.

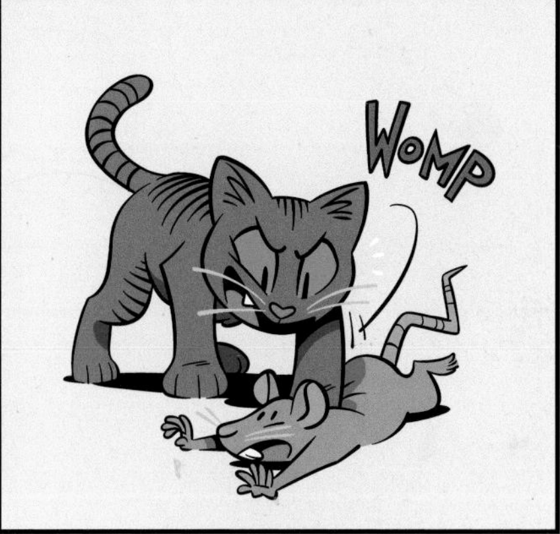

In time, these li'l fuzzballs will become the accomplished predators you expect!

Gosh, but this next part threw me for a loop.

Sorry! I thought I had that one.
ACK!

Leave me alone! I don't want any trouble, you hear?!
Kitten! That's no way to say hello!
ZIP

Wh-what is *she* doing here?
The same thing you are, kitten. There's room, so here she is.

But she's not a ***kitten!***
She was in our den, and that's where kittens go, so...?
B-but-but-but!

Family is a funny thing for cats like me. Litters all have the same mom, but different kittens may have ***different dads.***

Moms help each other care for kittens so they can trade off shifts guarding and hunting, making litters end up with ***multiple moms*** too.

And because cats are usually solitary, they expect any mammal in their nests to be, well, their kitten, and they'll treat them as such.

Lucky thing.

Hi.

Anyway, she showed up in the nest, and that's where kittens go.

Who was I to argue?

Sweet, right?

Sweet?! It's CRAZY! Cats are supposed to eat their prey, not befriend them!

What about our carnivore's physique?

Z
Z
Z
Z
What about our predator's senses?

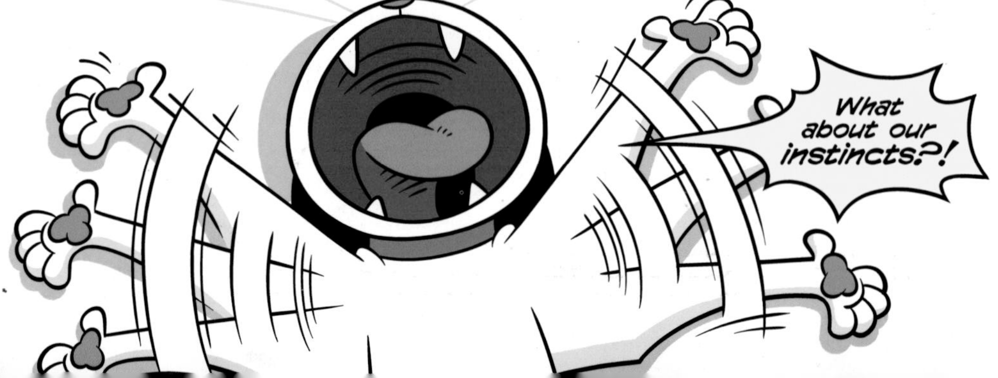
What about our instincts?!

It's a good question, and almost a century ago, a pair of experiments was run to answer it.
Kittens were raised in different situations like:
Alone.
With a rat-killing mom.
With kittens and rats.
With only a rat roommate.
A
By measuring the differences in predatory behavior between these groups, perhaps something could be learned about what's really *natural* for a cat.

Kittens raised with rat-hunting moms behaved how you'd expect: almost all of them were rat-killers themselves from a very young age.

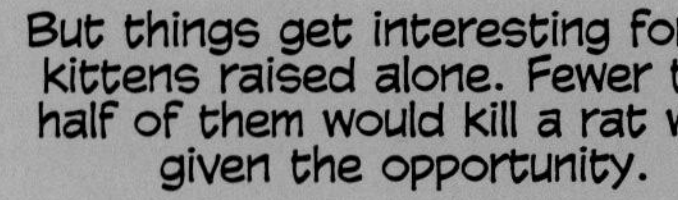

But things get interesting for the kittens raised alone. Fewer than half of them would kill a rat when given the opportunity.

?

Not a single kitten would kill the type of rat they were raised with.

Most astonishing of all, the kittens raised only with rats grew to, well, *love* their rat.

Make no mistake, cats can still figure out how to hunt without seeing their mom at work, though their choice of prey may differ.

But if a kitten, free from encouragement one way or another, is as likely to be friendly with their prey as not, are their instincts ***predatory*** or ***peaceful?***

A cat
has a cat's
body, with all
the teeth and
claws that
includes, so they
can only do cat
things.
That's a
big collection
of possible
actions to
choose from,
determined by
their physical
nature.
What
actions
a cat *does*
choose,
though...
...that
depends on
experience.
That
depends on
how they're
nurtured.

Is there any way to add or subtract from that collection of actions? To change a cat's nature?
Hmm... first we'd have to consider what ***determines*** that nature. The answer is something...

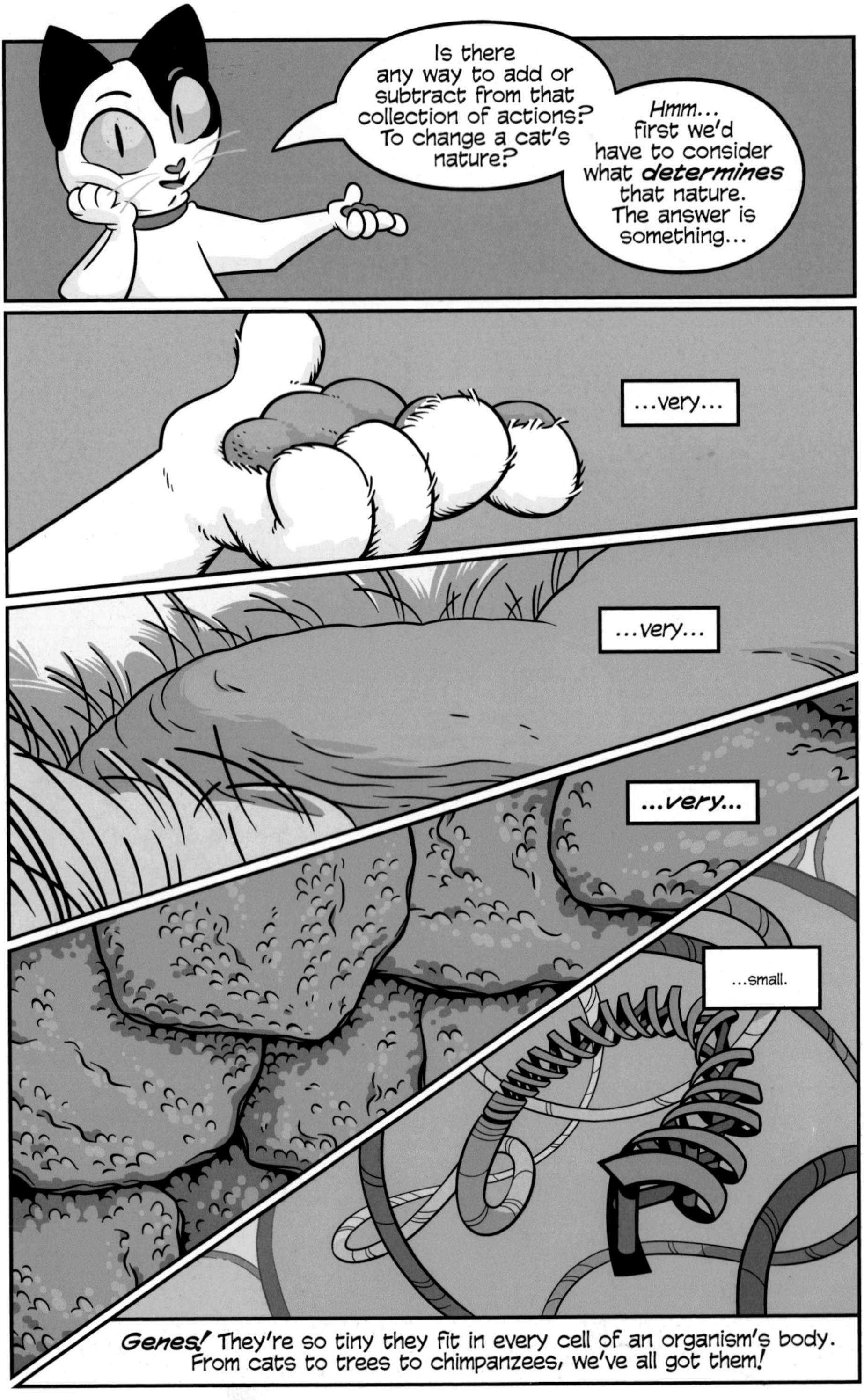

...very...

...very...

...very...

...small.

Genes! They're so tiny they fit in every cell of an organism's body. From cats to trees to chimpanzees, we've all got them!

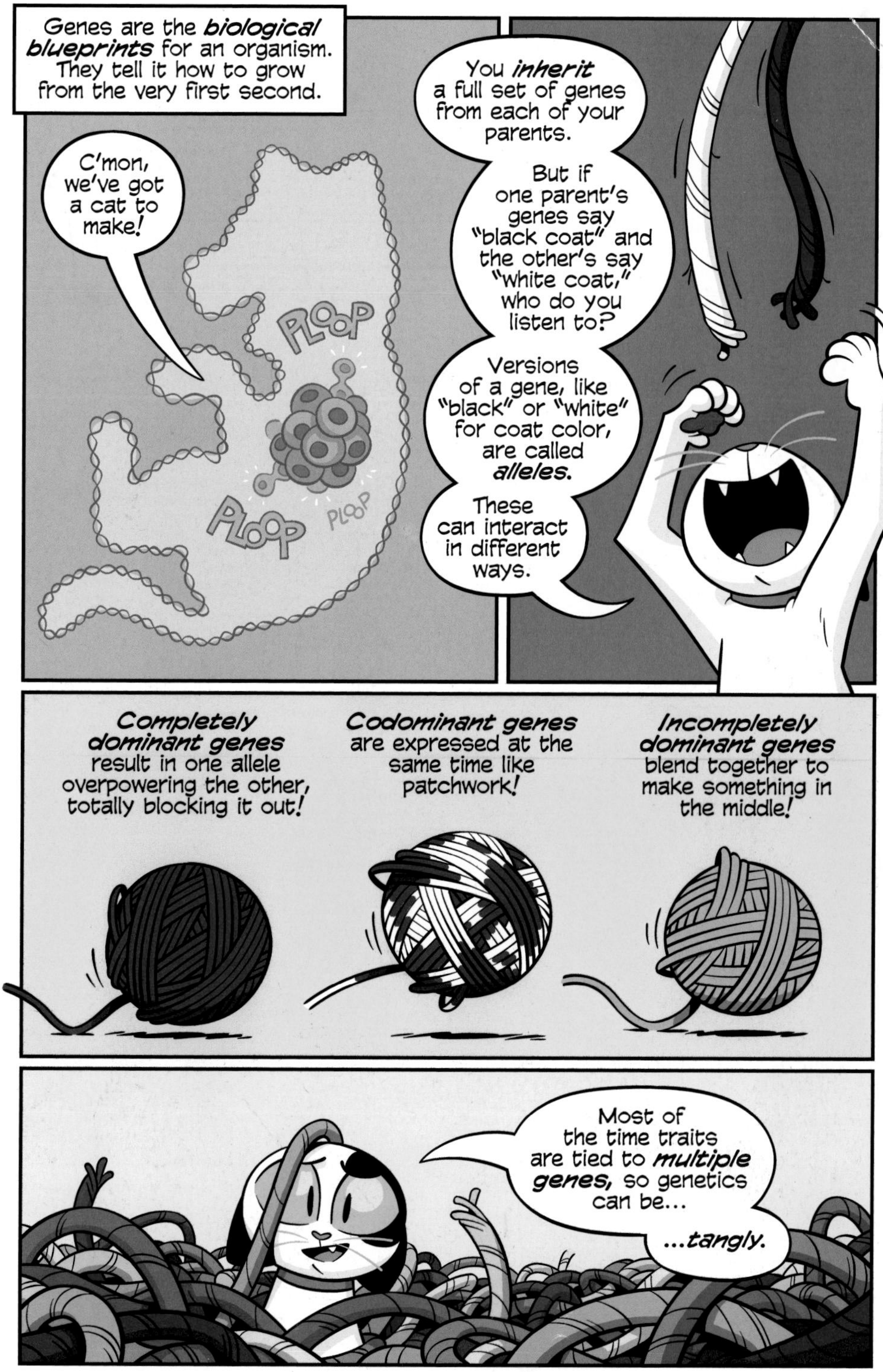

Genes are the ***biological blueprints*** for an organism. They tell it how to grow from the very first second.

C'mon, we've got a cat to make!

PLOOP

PLOOP

PLOOP

You ***inherit*** a full set of genes from each of your parents.

But if one parent's genes say "black coat" and the other's say "white coat," who do you listen to?

Versions of a gene, like "black" or "white" for coat color, are called ***alleles.***

These can interact in different ways.

Completely dominant genes result in one allele overpowering the other, totally blocking it out!

Codominant genes are expressed at the same time like patchwork!

Incompletely dominant genes blend together to make something in the middle!

Most of the time traits are tied to *multiple genes,* so genetics can be...
...*tangly.*

The content of your genes is your genotype, and the trait that you show, whether that's physical or behavioral, is your phenotype. Genotype influences phenotype, but individuals with very different genes might look and act very similarly. The opposite is true too!
Your genotype will never change, but your behavioral phenotype, well, we already know that your environment can change that.
Your environment is everything around you. From the rock you rest on to the river you drink from and from the climate you live in to the company you keep. Everything!

...when some of those traits *perform better* than others...

...and when those traits can be *passed down* from generation to generation, there is inevitably *natural selection*.

Over time, as individuals with helpful traits are successful enough in their environment to have more offspring than those with less helpful traits, natural selection makes those helpful traits ***more common***.

In this way, species change and *evolve* to best fit their environment!

For way more on genetics, check out ***Science Comics: Dogs!***

But natural selection can only act on *phenotypes,* right? If an organism's genes tend to make it shy toward large, upright, mostly bald mammals, but there aren't any of those in the organism's environment, that trait is kind of a *secret,* isn't it?
It won't show up as a phenotype, so it can't be rewarded *or* punished.
YAWN!
That means a species will have all sorts of varying *genotypes* related to this trait.
Some individuals will have it in their nature to be *shy,* others to be *bold,* most somewhere in between, but no one will *know* because there aren't any large, upright, mostly bald mammals around!
What if the environment changed? What if, oh goodness, *what if large, upright, mostly bald mammals showed up?*

Heya, kitties!
Just as I was settling in too.

Aw, hello, Simon!
Tilly, there's my girl!
You'll warm up one of these days, Socks!
LWS

Shoot, I keep asking them not to block your room.
All right, who wants some fresh water?

Oh. I haven't seen you before...

For some reason, I wasn't scared.
It's okay.

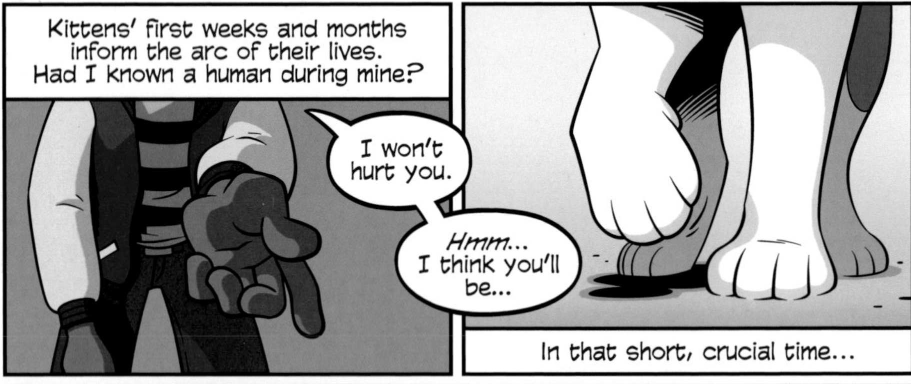
Kittens' first weeks and months inform the arc of their lives. Had I known a human during mine?
I won't hurt you.
Hmm... I think you'll be...
In that short, crucial time...

...had someone been kind?
...Bean.

Almost any kind of cat would've run away from him, but there's something different about cats like me. Something in the very nature of us ***domestic cats.***

10,000 years ago, on the other side of the world from where the very last saber-toothed cats were going extinct, there was one certain species of small cat.

Here in the Fertile Crescent, the ***North African wildcat*** lives much like so many cats have for so many years.

Like all wild cats, this one is an opportunistic hunter. Hungry or not, they'll try to get any meal they can. Who knows when the next one might come along?

Now there's a new sort of animal in this cat's environment—a large, upright, mostly bald mammal called a *human*. They've never had anything this cat wants, so the two have been happy to stay apart from one another.
waddle waddle

These humans used to be traveling hunters, but lately they've been staying in one place to farm grains. A steady source of food sure makes things easier for them.
NAB

GULP!

What's that strange aftertaste?
Is it... *grain?*
It makes things easier for *mice* too.

STOMP

It *is* grain!

Where're they getting this stuff?!
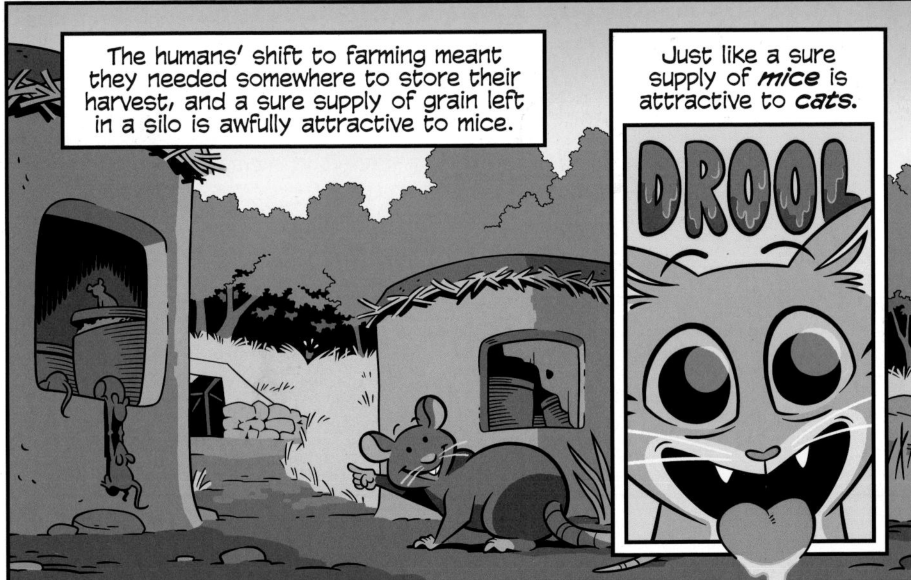
The humans' shift to farming meant they needed somewhere to store their harvest, and a sure supply of grain left in a silo is awfully attractive to mice.
Just like a sure supply of ***mice*** is attractive to ***cats.***
DROOL

Oh, but this wildcat just can't do it.

Perhaps a different wildcat will have it in them to go further.

What's the difference between these two cats? One was able to take advantage of this ***new environment***, but one wasn't. When it comes to humans, one is ***shy*** and the other ***bold***. It was hidden in their genotypes, and this game-changing difference in phenotype was never expressed until now!

Now that the bold phenotype is *advantageous,* that cat will have more opportunities to *pass down* the trait to further generations.
Their kittens, better-fed, healthier, and more likely to survive because of their boldness, will do the same!

Through natural selection, each generation's cats are more tolerant of humans than the one before. With more cats taking advantage of this concentrated food source, they'll also naturally become more tolerant of ***each other***.

By adapting to the human environment, this population of North African wildcats has taken its first step toward *domestication*.

Pre-domestic cats and humans had a ***commensal relationship,*** which means that one benefited without having much of an effect on the other. In this case, cats benefited from all the mice that humans attracted.

What did humans get? Cats may have provided them with some amount of ***pest control,*** but not very dependably. You could say that humans never hired us full-time but that cats showed up whenever we felt like it.

I'll get the next one. It's not like they're going anywhere.
ZZZ

Besides, humans' hardworking pet ***dogs*** had already been around for ***thousands of years*** and weren't bad mousers themselves.

Get outta here!

Were cats enough of an improvement to take over the job?

What ***new*** service could we bring to the table?

Take this seriously!
Eh, I've got other options.

It's the CUTE FACTOR!
Spread it far and wide: cats have been scientifically proven to be adorable.
We've got baby-like facial proportions that humans can't say no to.
Those big ol' eyes!
AWWW...
That round li'l head!

Those chubby cheeks!
AWWW...
That itty-bitty nose!
We're soft, fuzzy, and all-around pleasant to touch.
That tiny meow? Heart melting!
AWWW...
Who says there's no such job as simply being nice to have around?

Humans already had a habit of trying to tame us as ***cute li'l kittens...*** but now that they had mice we were willing to stick around as ***cute ol' cats.***

The North African wildcat is just one member of the *Felis silvestris* species. The others are:

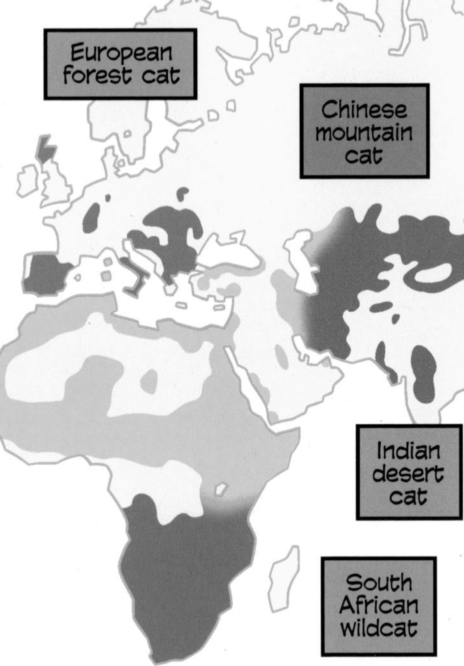

And wouldn't you know it? These regional flavors of *F. silvestris* can all have kittens with one another.

As pre-domestic North African wildcats started to migrate alongside humans, they continued to interbreed with the locals whenever possible. One paw was now in civilization, but the other remained firmly in the wild.

Cats found our biggest fans in Egypt, where we were kept as pets at least 4,500 years ago!
MIW
These folks went so ***cat crazy*** that we changed their gods! Take ***Bastet,*** who originated as the ferocious lion-headed protector of humankind.
As time passed she transformed to have the look and playfulness of a small cat!
Ancient Egyptians were the first to take complete control over cats' mating habits, breeding innumerable cats for religious rites.
I can stop whenever I want—promise! I just don't want them to get ***lonely!***
You could call this guy the first ***crazy cat person.***

Cats continued to spread across the world, often on ships.
On board, we were lucky charms who could stay out of the way while also catching a rodent or two.
Once we found ourselves in new environments, our original adaptations to life in North Africa weren't as helpful.
When environments change, so must living things. Slowly, and with no input from humans, pre-domestic cats began to differentiate from one another.
In warm southeast Asia, short hair and slender bodies kept us healthy and happy.
I'm on island time.
In cold northern Europe, long hair and stocky bodies were the best to have.
And I'm feeling cozy!
These and other distinct *natural breeds* developed wherever populations of cats were geographically isolated from one another.

By this point, cats have been hanging around humans for thousands of years.

Are we ready to put a label on it? Are cats ***domesticated?***

There are three essential criteria to meet.

First, we've got to have our food provided by humans.

Check!

But...

Second, we've got to be sheltered or have our movements decided by humans.

Check!

But...

Third, our breeding has got to be controlled by humans too.

Check!

But...

Hmm... Kinda?
Maybe?

It's undeniable that cats can get along in the human environment and that humans provide *some* of our needs.

It's debatable whether or not humans keep cats on a short enough leash to be truly, 100% domesticated, though.

If cats were simply *tame,* individuals would tolerate humans, but each generation would have to be tamed all over again. Nothing in the species' nature would change.

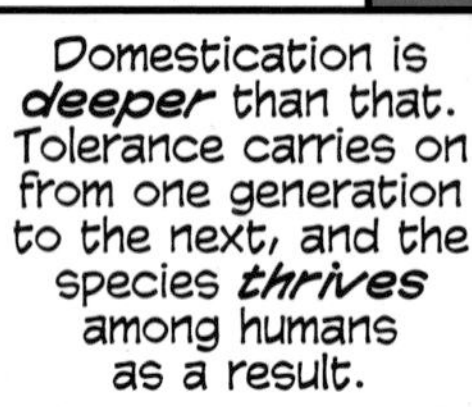

Domestication is *deeper* than that. Tolerance carries on from one generation to the next, and the species *thrives* among humans as a result.

Soon we arrived at this human's home.

Our home.

He'd thought of everything! My new environment had so much of what cats love in the wild.

But no wild cat could have been as content there as I was. They don't have it in their nature to appreciate my new *friend.*

That's not to say all domestic cats would appreciate him either.
HISS!
Our socialization period doesn't last forever, remember, and a domestic cat who isn't properly nurtured early on will likely always be wary of humans.

Still, our nature has changed enough that even the shyest domestic cat is more human-tolerant than the boldest wild cat.

Those super-remote species that go generations without meeting humans are sweet, trusting excejptions relative to other wild cats.

But you've got to be ***extremely*** bold to put up with ***this.***
Gosh, we're a cute pair.

Meows are actually rare between cats. Most meowing is directed at humans—maybe because they don't understand smells like other cats can.

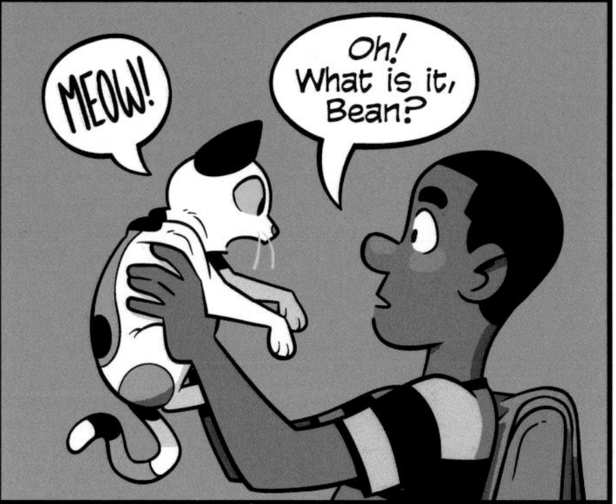

Oh! What is it, Bean?

We can make all sorts of meows:
MOWW
new
EEEOW
MAOW MAOW MAOW
MRROW
But they likely don't have fixed meanings between us cats.

So in a domestic setting, cats and humans have to agree on what meow means what.

MEOW!
Pats?

Hang on, he'll get it.
MEOW!
Food?

Ta-da! A secret language between each cat and their person*!*

Let's move on to Meow #2, huh?

Okay, *"MEOW #1 = FOOD."* Got it.

Domestic animals like dogs have lots of different jobs, so humans want kinds of dogs specifically suited to each one.

Huge mountain dogs to guard livestock...

...short sausage dogs to tunnel through burrows...

...stout, scrappy dogs for pest control...

...their *forms* fit their *functions.*

What function might cats have that our hunter's bodies don't already suit?

I seem to remember hearing something about being cute...

Early domestic cats were all brown with ***mackerel tabby*** patterning, just like their wild ancestors.

Changing coat colors often follow domestication. Just look at cows, pigs, or dogs. They're spotty and splotchy, and they come in every shade.

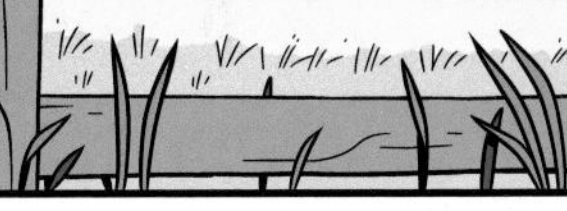

One reason for this may be that camouflaging coats don't offer any benefit outside of the wild. If you aren't sneaking up on prey or hiding from predators, why be sneaky?

And looking at it from the other direction, cats without camouflaging coats are unlikely to survive long enough in the wild to pass those genes on.

One specific mutation makes a developing cat's coat temperature-sensitive, causing the fur on their cooler extremities to be darker or lighter than the fur on their warm core.

Say, that's a pretty good look! With humans here to help, these genes are able to stick around.

If a mutation strikes humans' fancy, they'll want the cat that has it to pass the trait along to as many kittens as possible.

Artificial selection, different from natural selection, is when *humans* breed species based on *humans'* priorities, not environmental fitness. It means humans call all the shots!
I must have more!
You two: kiss!

Favored mutations can be *very* persistent, and a little bit of human intervention goes a long way. For instance and for obvious reasons, Vikings' favorite cats were *orange tabbies*.

They loved the little fuzzballs and brought them on ships during raids.
For glory!
For gold!
For kitties!

Those adorable mutants worked their way into the local populations, and while the Vikings are long gone, orange tabbies are *still* more common where they once were!

By strictly controlling which cats mate with which, humans choose the traits kittens are likely to show.
Artificial selection can be applied to any trait humans fancy and, for cats, that has almost always been our *looks.*
Over time and with enough discipline, distinct *breeds* can be created.
These are types of cats that have only mated within their own breed for generations and meet strict physical criteria.

Today there are almost 40 genetically distinct breeds and many variations of each.
Here, kitty-kitty!
Russian Blue
Manx
Sphynx
Cornish Rex
Abyssinian
American Curl

Oriental
British Shorthair
Balinese
Scottish Fold
Maine Coon
Egyptian Mau

Persian and ***Siamese*** are two of the oldest recognized breeds. Look at how different they've become thanks to artificial selection!

Beyond their physical differences, these two are personality opposites.

Persians tend to be quiet, lazy, and uninterested in hunting.

Siamese tend to be loud, hyper, and the stuff of prey's nightmares.

Differences as large and predictable as these must have been of ***some*** interest to breeders over the years, even if they were secondary to looks.

How might being active and vocal have become so closely tied to the Siamese breed? Perhaps talkative, playful cats were the most prized throughout their notably long history among royalty.

Meanwhile, a longhair Persian who needs lots of grooming gains health benefits from being calm enough for brushing. By selecting for the longhair ***look,*** humans unintentionally also select the calm ***personality.***

Domestic cats are even able to interbreed with some species of wild cats beyond *Felis silvestris*. This is how you end up with cats like the Bengal, Safari, and Savannah.

Cross-species breeding seems to knock cats a few steps down the domestication track, making crosses far less suitable as pets. That's not surprising since the wild parent has ***none*** of the genes for human tolerance.

But by mating hybrids with domestic cats over several generations, that friendly domestic temperament can be restored—now with a brand-new look!

Cat domesticity is fragile, surprisingly or not. It seems easy for our species to revert to its wild ways. In fact, the way humans keep cats today is next to incompatible with the strict rules other animals follow.
Are humans really *essential* for shelter if *15%* of house cats leave a home within their first year there?
Are they *essential* for food when many cats are left outside to hunt local wildlife?
How can humans select for domestic genes when not even *3%* of kittens come from planned mating and over *30%* are adopted as strays with unknown origins?

More domestic cats are born in the United States each day than there are lions in the wild, and a big share of those are strays.

In order to improve cats' well-being by preventing overpopulation, humans often turn to ***trap-neuter-release (TNR)*** programs.

TNR involves catching stray cats, giving them minor surgery to prevent them from having more kittens...

...and releasing them back into their territories. With hard work and planning, this can be a big help.

But! The cats that tend to get trapped are the least cautious individuals—the same that are least afraid of humans. That means wary cats make up a bigger part of the gene pool. Are humans accidentally selecting for shy cats? Are they ***undomesticating*** cats?

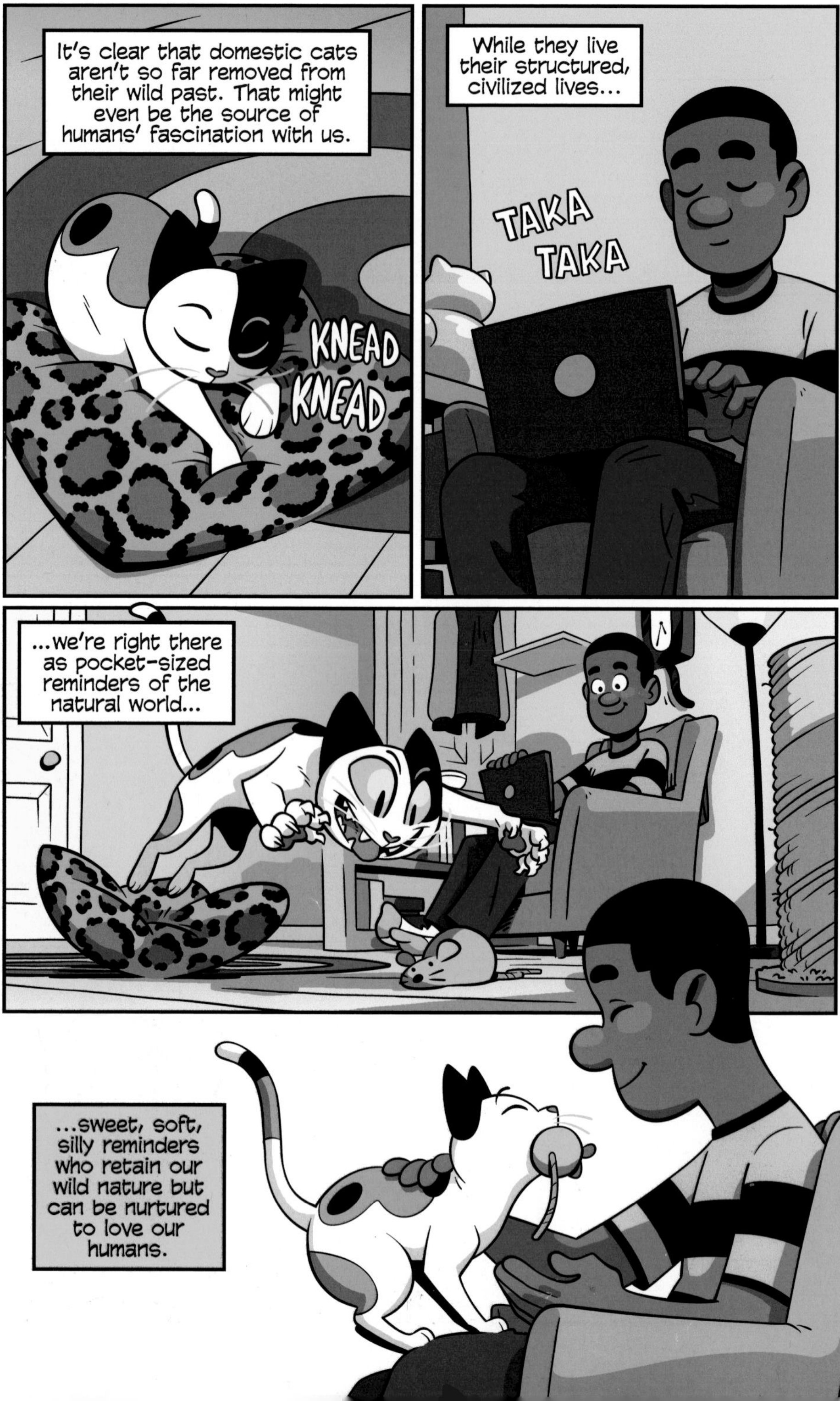

It's clear that domestic cats aren't so far removed from their wild past. That might even be the source of humans' fascination with us.
KNEAD
KNEAD

While they live their structured, civilized lives...
TAKA
TAKA

...we're right there as pocket-sized reminders of the natural world...

...sweet, soft, silly reminders who retain our wild nature but can be nurtured to love our humans.

What? Of course this is all relevent to your question!
It's really sweet, besides!
ON FIRE OR OFF TOPIC?

Jeez.

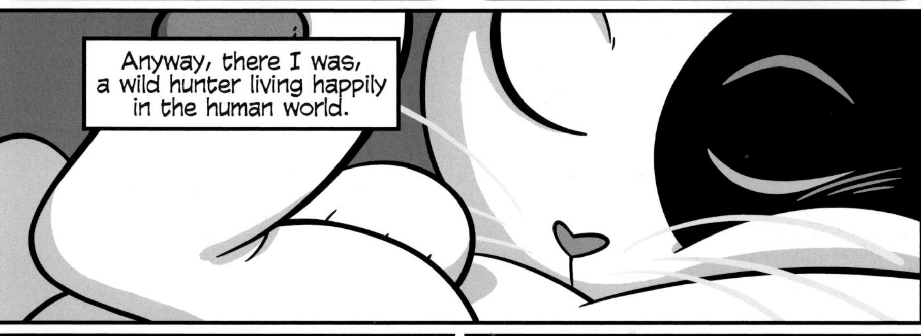
Anyway, there I was, a wild hunter living happily in the human world.

What a contrast!
!

I was just being me, a li'l lion, a carnivorous hunting machine, an unstoppable survivor...
Haha! Bean, what are you even *doing?*
TAP

301
2,326
79,338,228
Trophy cat napping
2,013,323,484 views
First!
OMG SOOOOOOOOO CUTE :D
Such a sweet kitty!!!
lol
Anyone know this cat's name?!
<3 <3 <3
Me during class
fake.
844,779,

Breathtaking. It's only natural to admire such a perfect specimen of cathood.

Got everything you need, then?
Good.

Cut!
Great work, everyone, now pack it up!
Bean! Come on, girl, time to buckle up for landing.
Yeah, I love you too.
Meow!
SKRRRRRT

CATS! CA
TS!
TROPHY CAT CONFIRMS:

NAP CHAMP
"CATS GOTTA BE CATS"

Meow.

Meow
meow.

Rowr?
Maow maow
maow!

Mew!

Sigh...
Whose bright idea was it to interview a *cat?*
STAR Tales
EDITOR

—GLOSSARY—

Bunting

A common behavior in which a cat affectionately butts or rubs its head against another animal or object to deposit its scent.

Domestication

The process of taming a species to be kept as a pet or a work animal, which usually creates a dependency so that the animal loses its ability to live in the wild.

Environment

An organism's surroundings, including everything from the climate to other plants and animals.

Evolution

Changes in a species' traits and genes over time, allowing them to adapt to and diversify within their environment based on:

Natural selection: The process by which the most fit organisms adapt to their environment, survive, thrive, and reproduce at higher rates than those less fit.

Artificial selection: The process by which humans intentionally breed organisms to display desirable traits in their offspring.

Flehmen

A cat's scrunched-nose, opened-mouth facial response to pheremones. This engages the vomeronasal organ, a secondary scent organ located above the roof of the mouth.

Gene

A segment of a DNA molecule that serves as the basic unit of heredity. Genes control the characteristics that an offspring will have.

Genotype

The genetic makeup of an organism. It can contain code for traits that are not displayed.

Hypercarnivore

An animal whose diet consists of at least 70% meat.

Matriarchy

A social organization led by females. For example, domestic cat colonies have a stable group of females who raise kittens while males wander in and out.

Melanism

The development of the pigment melanin resulting in a black coat. Melanistic individuals are somewhat common in jaguars and leopards and are sometimes referred to as "black panthers." Melanism has also been seen in some small species.

Midden

Nature's litter box. These are usually found along travel routes or territorial boundaries, and any given midden will be reused until unhealthy parasites build up in it.

Phenotype

The observable features or characteristics of an organism, such as appearance and behavior.

Righting reflex

The series of motions that allows a cat to reorient itself during a fall to reduce the risk of injury.

—GLOSSARY CONTINUED—

Saber-toothed cat
Any of a number of extinct cat species with notably elongated canine teeth. These include *Smilodon*, *Machairodus*, *Homotherium*, and others.

Saccade
A rapid eye movement that helps a cat precisely track its prey.

Scruff response
Important during kittenhood, this causes a cat to reflexively go limp when held by the loose skin on the back of its neck.

Tail up
A friendly posture expressed by social cat species such as lions and domestics.

Tapetum lucidum
A tissue behind the eye's photoreceptors that reflects light past them a second time to improve night vision. This reflection causes cats' eyes to appear to glow in the dark.

Trap-neuter-release
A strategy to reduce the stray cat population by surgically preventing the cats from reproducing.

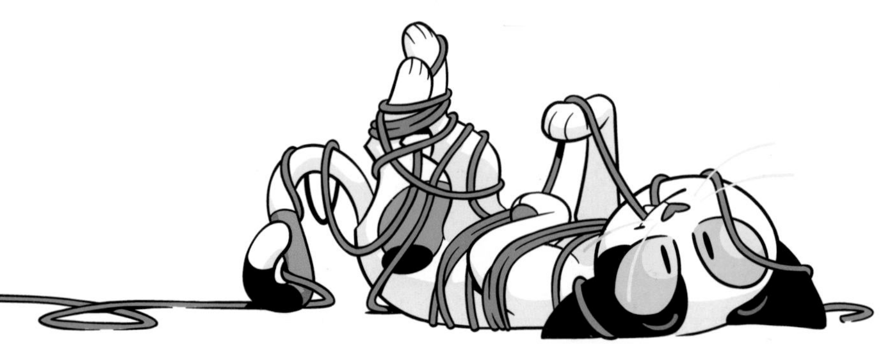

—BIBLIOGRAPHY—

Allen, William L., et al. "Why the Leopard Got Its Spots: Relating Pattern Development to Ecology in Felids." *Proceedings of the Royal Society B*, vol. 278, 2010.

Banks, Martin S., et al. "Why Do Animal Eyes Have Pupils of Different Shapes?" *Science Advances*, vol. 1, no. 7, 2015.

Bradshaw, John. *Cat Sense: How the New Feline Science Can Make You a Better Friend to Your Pet.* Basic Books, 2013.

Francis, Richard C. *Domesticated: Evolution in a Man-Made World.* W. W. Norton & Company, 2015.

Hart, Benjamin L., and Lynette A. Hart. *Your Ideal Cat: Insights into Breed and Gender Differences in Cat Behavior.* Purdue University Press, 2013.

Kuo, Zing Yang. "The Genesis of the Cat's Responses to the Rat." *Comparative Psychology*, vol. 11, no. 1, 1930.

Kuo, Zing Yang. "Further Study on the Behavior of the Cat Toward the Rat." *Comparative Psychology*, vol. 25, no. 1, 1938.

McNamee, Thomas. *The Inner Life of Cats: The Science and Secrets of Our Mysterious Feline Companions.* Hachette Books, 2017.

Ottoni, Claudio, et al. "The Palaeogenetics of Cat Dispersal in the Ancient World." *Nature Ecology & Evolution*, vol. 1, no. 139, 2017.

Sunquist, Mel, and Fiona Sunquist. *Wild Cats of the World.* University of Chicago Press, 2002.

Tucker, Abigail. *The Lion in the Living Room: How Cats Tamed Us and Took Over the World.* Simon & Schuster, 2016.

Turner, Alan. *The Big Cats and Their Fossil Relatives.* Columbia University Press, 1997.

Turner, Dennis C., and Patrick Bateson, editors. *The Domestic Cat: The Biology of Its Behaviour.* 3rd ed., Cambridge University Press, 2014.